Hoofed Mammals of British Columbia

ROYAL BC MUSEUM HANDBOOK

HOOFED MAMMALS
OF BRITISH COLUMBIA

David Shackleton

Volume 3
The Mammals of British Columbia

REVISED EDITION

Published by the Royal BC Museum, 675 Belleville Street,
Victoria, British Columbia, V8W 9W2, Canada.

Printed in Canada.

MIX
Paper from
responsible sources
FSC® C016245

Library and Archives Canada Cataloguing in Publication

Shackleton, David M., 1944–
Hoofed mammals of British Columbia / David Shackleton.
– Rev. ed.

(Royal BC Museum Handbook ; ISSN 1188-5114)
Includes bibliographical references and an index.
ISBN 978-0-7726-6638-3

1. Ungulates – British Columbia. 2. Ungulates.
I. Royal BC Museum II. Title. III. Series: Royal BC Museum handbook.

QL737 U4 S52 2013 599.609711 C2013-980023-9

CONTENTS

PREFACE

This is the third in a series of six handbooks revising the Royal BC Museum's Handbook 11, *The Mammals of British Columbia* (Cowan and Guiguet 1965), which is no longer in print. Like the other volumes in the series, this handbook emphasizes natural history, distribution and identification, using recently published information.

Changes have occurred in taxonomy and nomenclature since the publication of Cowan and Guiguet's handbook. The development of new genetic techniques allows biologists to examine evolutionary relationships through comparisons of DNA. As a result of these developments, we can expect further taxonomic revisions not only of hoofed mammals, but for most species.

The technical name for a hoofed mammal is *ungulate*, and I use both terms throughout the book. To avoid unnecessary confusion for new readers, rather than use terms for the different age-sex classes of hoofed mammals that are often species-specific (e.g. buck, doe, fawn, nanny, billy, kid, ram, ewe, bull, cow, calf, lamb), I simply refer to them as males, females, adults, juveniles and young of the year.

My objectives in the General Biology section are to provide background information on the biology and ecology of hoofed mammals, and to encourage the reader to ask questions, and even make predictions, about why these mammals look and behave the way they do. Many hoofed mammals have distinct markings or particular kinds of weapons, and I hope that you will begin wondering about how an animal might use these markings or how it might employ its weapons to fight or display. The Species Accounts, which give details about the natural history of each of BC's hoofed mammals, is accompanied by the excellent drawings of the live animals by Michael Hames and of their skulls by Denise Koshowski. There is still much to be learned about BC's ungulates and I hope this book will encourage further work on these fascinating animals.

In this revised edition, I have corrected errors in the first, made some revisions, including updating population estimates and conservation status, and added new references. The most significant revision was combining California and Rocky Mountain Bighorn Sheep into one account – recent taxonomic studies suggested no difference between these former subspecies, and provincial authorities have officially adopted this reclassification. I have also inserted new images wherever possible and added a colour section that highlights interesting characteristcs of these animals.

GENERAL BIOLOGY

What is a Hoofed Mammal?

Hoofed mammals are a group of terrestrial animals characterized by their feet, or more specifically by features called hooves (figure 1). Hooves are the outer coverings or sheaths surrounding the last bone of the toe, and are made of keratinized epidermis or skin, the same material as our fingernails. This tough sheath is called the *unguis*, from the Latin for nail or claw, and the technical name for a hoofed mammal is "ungulate", derived from the Latin *ungula*, meaning hoof. Immediately under the hoof is the softer sub-unguis, and behind this is the pad. All mammals with keratinized plates, such as nails or claws, are placed together in a broader division of eutherian mammals called the Ferungulata. Besides the ungulates, this division contains animals such as the carnivores and the primates, and so includes humans.

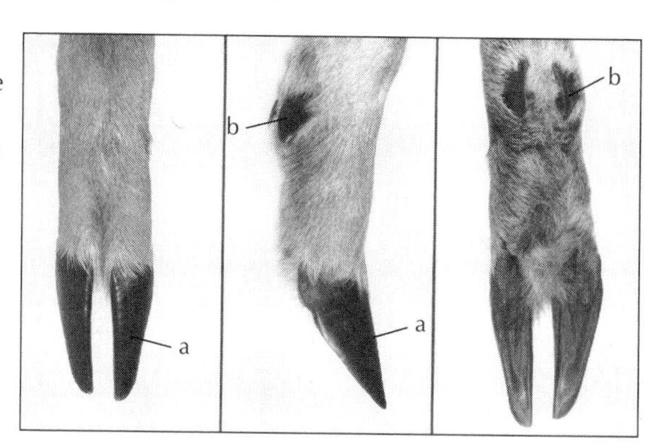

Figure 1. Front, side and rear views of a yearling Columbian Black-tailed Deer's foot to show the main hooves (a) and the dew claws or lateral hooves (b).

Ungulates are an extremely successful group. They are the most abundant large land mammal with about 180 species world wide. They are native to all continents, except for Antarctica and Australia. Most ungulates have a common general ecological strategy, are herbivorous and share some specialized morphological adaptations. Their diet probably explains much of their success because plants provide such a diverse and abundant food source. Besides familiar wild species like deer and Bighorn Sheep, many domestic animals such as horses, cattle, pigs, sheep and goats are also ungulates. These domestic forms play a crucial role in agriculture and food production, and together with their wild relatives, are extremely important to humans throughout the world.

The Evolution of Ungulates

The ancestral ungulates probably arose from a group of early mammals called Condylarths that lived in the early Palaeocene epoch (table 1). Later in the Palaeocene, the Paenungulates or so-called primitive ungulates, diverged from the Condylarths. These sub-ungulates persist to this day as an interesting grouping of three orders: elephants (Proboscidea) in Africa and Asia, the diminutive hyraxes (Hyracoidea) of Africa, and the aquatic dugongs and manatees (Sirenia).

The true ungulates are divided into two orders, based primarily on their foot structure and number of toes (see Form and Structure). Perissodactyl ungulates have an odd number of toes (one or three, although Tapirs have four toes on their forefeet), while Artiodactyls have an even number of toes (two or four). Modern Artiodactyls are the most common and diverse of the two orders, comprising nine families: swine, peccaries, hippopotamuses, camels, mouse deer, deer, giraffes, the pronghorn and horned ungulates (bovids). Today, only three families of Perissodactyls remain – horses, rhinoceroses and tapirs. All are herbivores, but unlike Artiodactyls, modern odd-toed ungulates have a single stomach and process their food quite differently (see Dietary Habits).

Recent molecular and morphological evidence has led some researchers to suggest that Artiodactyls, specifically hippopotamuses, are the closest relatives of Cetaceans (whales). They have suggested that the two orders be combined into one or be placed in a new superorder, called Cetartiodactyla in either case. For the purposes of this book, I will continue to use Artiodactyla, because it refers only to terrestrial members of the order.

Table 1. Geological time scale covering the history of ungulates and other mammals. (After Macdonald 1984 and Vaughan 1986.)

Era	Period	Epoch	Start of Epoch (years before present)	Major ungulate and other mammal events
Cenozoic	Quaternary	Holocene or Recent	10,000	Epoch since the end of the Ice Age. Extinction of many large mammals, including some ungulates, occurs 10,000 to 12,000 years ago.
		Pleistocene	1.6 - 2 million	Many ungulates, along with other animals and plants, migrated between continents across land bridges.
	Tertiary	Pliocene	7 million	All modern families of ungulates present by the beginning of this epoch.
		Miocene	26 million	Grasses flourish, followed by the emergence of the first specialized grazing ungulates.
		Oligicene	38 million	The ungulates diversify.
		Eocene	54 million	Ancestors of Artiodactyls and Perissodactyls, along with other modern mammalian orders appear.
		Palaeocene	65 million	Condylarths, probable ancestor of ungulates, appear, followed later by the first primitive ungulates.
Mesozoic	Cretaceous		130 million	Appearance of the first marsupial and placental mammals.
	Jurassic		190 million	Archaic mammals begin to diversify.
	Triassic		230 million	The first mammals appear.
Palaeozoic	Permian		280 million	Therapsids, the ancestors of mammals, appear.

Artiodactyls and Perissodactyls first appear in the fossil record in the early Eocene. At that time, the climate was much warmer than today and tropical forests grew as far north as the Canadian Arctic. The Perissodactyls were initially more successful, quickly evolving many different species that ranged in size from 5 kg to more than 1,000 kg. Most were browsers, and these early Perissodactyls were probably the first ungulates with an almost wholly herbivorous diet. Artiodactyls at this time were still small, usually less than 5 kg, and most retained the primitive, more omnivorous diets of their ancestors. But by the end of the Eocene, the world's climate had changed once more. There were now definite seasons in the northern and southern hemispheres, and these caused significant but predictable variation in the growth and abundance of plants.

Under these new conditions, plants evolved rapidly and herbivores were swift to respond. Artiodactyls especially began to diversify. Many large species appeared and all but the pigs and peccaries adopted an almost totally herbivorous diet. When grasses first flourished about 20 million years ago in the Miocene, much of the forest began converting to open savannah and the first specialized grazing ungulates appeared. By the end of the Miocene, all modern ungulate families were present. Modern Perissodactyls evolved first in North America, as did some Artiodactyls such as pronghorns and probably camels. By the late Pliocene, horses and camels had spread to South America and Eurasia.

The Pleistocene or Ice Age was a time of tremendous climatic fluctuation. Most present-day ungulate species, or their immediate ancestors, were roaming the earth by the middle of this epoch. During the many cold glacial events of the Pleistocene, ice masses accumulated on land around the world, forming huge glaciers and ice caps. During the last glacial event (Wisconsin), most of British Columbia was covered by ice up to two kilometres thick in some parts. With such vast amounts of water stored as ice around the world, sea levels fell and exposed the sea-bed around continents and islands. Some of this new dry land formed connections – land bridges – that allowed animals and plants to move between continents. One such bridge, the Bering Land Bridge or Beringia, connected Alaska and southeast Siberia. Several Artiodactyls, such as deer, Bison, wild sheep and Mountain Goats, along with other large mammals, migrated via Beringia from Eurasia to the New World, although few, if any, moved the other way at this time, probably because the way south was blocked by ice. Much earlier, before the Pleistocene, horses and camels had migrated from North America into Eurasia.

Partly as a result of the eastward migration during the Pleistocene, British Columbia was once home to many more species of ungulates and other large mammals than today. The Columbian and Woolly mammoths, American Mastodon, and Helmeted Muskox roamed southern Vancouver Island, while the Jefferson's Ground Sloth, Woolly Mammoth and Giant Bison inhabited other parts of the province. Then about 12,000 to 9,000 years ago, these and many other species of large land mammals became extinct in the northern hemisphere. Horses, wild asses and camels disappeared from BC and elsewhere in North America. Other mammals, such as Woolly Mammoth and Jefferson's Ground Sloth, vanished from the Earth altogether, although dwarf Woolly Mammoths survived until about 3,800 years ago on Wrangel Island off the coast of northeastern Siberia. As Peter Ward (1977) points out, this means that mammoths were still alive when the pyramids were being built in Egypt. But the mass extinction of large terrestrial mammals is almost as uncertain as the disappearance of the dinosaurs 66 million years ago. Scientists suggest that the evidence for the demise of the large mammals at the end of the Pleistocene points to some combination of climate change and efficient hunters as the most likely explanation. Colonizing humans have certainly caused the extinction of large animals in several areas around the world.

Today, Canada is left with 11 native ungulate species, and all but two of them, Muskox and Pronghorn, live in BC. With 9 wild native species and 17 subspecies, British Columbia has the greatest diversity of ungulates in Canada. The next most diverse province is Alberta, which also has 9 species but only 12 subspecies. Besides our native ungulates, BC is also home to feral populations of horses, sheep and goats, and an introduced wild species, Fallow Deer. In addition to these exotic ungulates, BC's wild species have been introduced to areas outside their normal range, including Sitka Black-Tailed Deer and Rocky Mountain Elk to Haida Gwaii. While legislation for game ranching does not generally permit native ungulates or closely related species to be farmed, it is legal to raise Fallow Deer, Bison and Reindeer (the domesticated form of Eurasian Caribou). The intent of these restrictions is to prevent hybridization with native ungulates. When introduced species escape, as they have, they can introduce new diseases, compete with native wildlife, damage habitat and be difficult to eradicate.

Selected References: Cannings and Cannings 1996, Guthrie 1990, Harington 1996, Kurtén and Anderson 1980, Macdonald 1984, Naughton 2012, Nowak 2003, Parker 1990, Pielou 1991, Price et al. 2005, Sutcliffe 1985, Vaughan et al. 2011, Ward 1997.

Form and Structure

Though not of a size to rival the giant dinosaurs, the largest land mammals that ever lived were ungulates. Members of the genus *Paraceratherium* [*Indricotherium*], Perissodactyls belonging to the rhinoceros family, roamed central Asia about 35 million years ago. Many of these immense creatures stood more than five metres at shoulder, had a skull longer than a metre and could browse tall vegetation to a height of eight metres. Some are estimated to have weighed more than 18,000 kg – about four times heavier than the largest living elephant. Today, the Hippopotamus and some of the rhinoceroses are the largest living ungulates, weighing over 3,000 kg, while at the other end of the scale are the mouse deer (Tragulids), the smallest of which weighs less than a kilogram. These diminutive hoofed mammals inhabit the dense tropical forests of west and central Africa, and southern Asia from India and Sri Lanka to Indonesia.

Giant forms of many different mammals appear frequently throughout evolutionary history and were relatively common even as recently as the Pleistocene. Probably the most well known giant ungulate was the Irish Elk, a large relative of Fallow Deer, whose huge antlers, weighing up to 45 kg, could spread almost four metres from tip to tip. But this was probably not the largest deer in terms of body weight. The Alaskan Moose, found today in northwestern British Columbia, is believed to be one of the heaviest deer ever to have lived, with some adult males weighing over 800 kg.

Ungulates share a set of important and unique adaptations that help them gather and process food, interact with each other for the ultimate goal of reproduction, and avoid being eaten – three fundamental life requirements for any animal. Specialized teeth and digestive systems help them maximize the nutritional value of their food, and sometimes body shape helps them feed more effectively (e.g., the Giraffe). Most ungulate species are social; they have developed a wide range of communication systems, as well as specific mating strategies and fighting styles. Avoiding predators has resulted in a variety of adaptations: long limbs for speed, a camouflaged coat and eyes positioned to give a wide field of vision. Some of the more important adaptations of ungulates are highlighted in the following sections.

Selected References: Macdonald 1984, Nowak 2003, Vaughan et al. 2011.

Adaptations for Survival

Because there is quite a lot of published information about hoofed mammals in general, I have organized this section under functional headings that relate to the major adaptations of morphology, physiology, behaviour and ecology. Most of what we know about their adaptations apply to ungulates in general, not just to species in British Columbia.

Staying Alive – Avoiding Predators

Unfortunately for ungulates, they are the most common prey of large carnivores (figure 2), so to survive they obviously need ways to avoid being eaten. If they are small, they can hide in thick vegetation, and may also evolve coat coloration and patterns that help camouflage them. For large ungulates and other species living in open areas, this is not an option. Instead, they must rely primarily on the benefits of group living, good vision and hearing to detect predators (see colour photograph C-5), and fleetness of foot to outrun them, because only a few ungulate species use cooperative defence tactics.

A prey species' first line of defence is usually to see the predator before it gets close enough to attack, and an ungulate's vision is adapted to detect movement over a wide field of view. While their basic eye structure is similar to most other mammals, there are significant differences. First, unlike ours, their eyes lack a central focusing spot (fovea), so they are not particularly good at focusing on fine details. Second, the iris and pupil are elliptical and horizontal, rather than round like ours, possibly providing a wide field of view. Third, and most important, their eyes are located on the sides of their head rather than at the front. Eyes at the front create the stereoscopic (binocular) vision and excellent depth perception that carnivores need to judge distance when chasing prey, or primates need when jumping from branch to branch. Eyes on the sides of its head, equipped with elliptical pupils, give the ungulate an almost 360° field of view; some species can probably see almost as well

Figure 2. Ungulates are the main prey for large predators, such as Cougars.

Figure 3. One of the main strategies ungulates use to avoid predators is to outrun them. A female Rocky Mountain Bighorn Sheep and her young flee from a Coyote.

behind as in front or to the side. While not good at distinguishing details, ungulates can detect movements over a wide area – a great advantage for spotting predators. Research on colour vision is limited and results uncertain. Based on the internal structure of the eye, it is likely that ungulates may be able to detect some colours, but their colour vision is not as well developed as in humans or some other species. Such a panoramic view of the world is difficult for us to imagine, but serves the ungulate well.

No matter how vigilant the prey, predators will sometimes get close enough to attack. So hoofed mammals need to be able to run fast. The evolutionary history of ungulates shows a gradual change in their limb structure toward fast running (cursorial locomotion), often over hard ground (figure 3). Most ungulates have evolved several adaptations to help them do this, including longer, lighter legs, small feet and specialized limb joints. Long legs are an obvious advantage for running fast, and most ungulates have long limbs. Longer legs mean both a longer lever action and a greater stride length. Most of the increased leg length is gained through elongated metapodials – the bones that lie between the wrist/ankle and the fingers/toes (figure 4). Speed can also be gained through lighter limbs that require less effort to overcome the inertia of locomotion. In hoofed mammals, lightness has been achieved by reducing musculature in the lower extremities

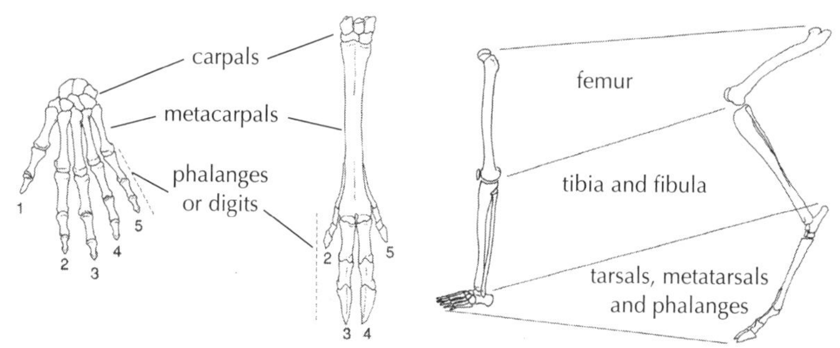

Figure 4. The pair of drawings on the left compares the bones in a human hand with those in a lower foreleg of a deer, a typical example of an Artiodactyl ungulate. Most ungulates have lost at least some of the bones of the foot. In this example, the first metacarpal and first digit have been lost, and the second and fifth are greatly reduced. The pair of drawings on the right shows the relative proportions of the elements of a human leg and a typical ungulate hind limb. It illustrates the elongation of the ungulate limb that evolved to increase running speed. The bones of the ungulate's foot, particularly the metatarsals, account for much of this elongation.

– replacing them with tendons and ligaments, and by reducing foot size. As mentioned earlier, the ungulate's specialized foot is one of its defining characteristics.

Like other vertebrates, ancestral ungulates had the basic plan of five digits (fingers or toes), but during their evolution to modern ungulates, they lost one to four outer toes. At the same time, the associated outer metapodial bones were lost or became very much reduced. In many modern species, the two central metapodials are also usually fused into a single functional bone, as in Columbian Black-tailed Deer or Rocky Mountain Bighorn Sheep (figure 4). The number of remaining functional toes is what we use to classify them as either Perissodactyls or Artiodactyls. The feet of all wild species of Artiodactyl living in British Columbia have lost the first toe and three metapodials. What remains are the two central metapodials fused into a single unit, two functioning toes (the third and fourth), and two greatly reduced outer toes (the second and fifth) called lateral hooves or dew claws (figures 1 and 4). Both of these small outer toes are also covered by a hoof, but we do not know their function in most species. The most extreme example of foot reduction is in the modern horse family. Zebras, asses and horses have retained only the third metapodial, the cannon bone, and the associated single toe, although vestigial metapodials or splint bones may occur. At the other extreme, heavy-bodied species such as rhinoceros and hippopotamus have rather short, splayed digits that form the wide foot needed to support their weight.

Hoofed mammals also gain running efficiency and speed through modifications of the leg joints and the way that the limbs are attached to the vertebral column. Many of the limb joints restrict their legs to strong, forward and backward motions. This allows the ungulate to transfer power more effectively to forward movement. The hind legs are firmly attached to the sacral section of the vertebral column via the pelvis. This strong attachment is important because the hind limbs provide most of the forward thrust, so the force is transferred directly to the body through the backbone.

Early forms of hoofed mammals may have lived in forested and moist habitats with soft ground, but most modern species inhabit open areas where the ground is hard. An animal running for its life with a predator in pursuit receives a terrific impact on its body every time the front feet hit the hard ground. But because, as in most mammals, the ungulate's scapula (shoulder blade) is not attached to the backbone and floats somewhat loosely, there is enough flexibility to cushion the impact of the front limbs striking the ground. The flexibility of the front limbs is even greater because, like other cursorial animals, ungulates have lost the clavicle (collar bone). As a result, the scapulas can move more freely when the forelimbs swing forward, increasing stride length and speed. In Artiodactyls, the astragalus bone in the hind ankles has evolved to form an efficient double-pulley system giving the lower hind limbs even more flexibility and springiness.

As a result of all the changes to the lower limb and foot, most modern ungulates walk with a type of gait called unguligrade, which improves the animal's ability to move quickly over a hard surface. The unguligrade gait, in which the animal's weight is on keratin hooves surrounding toe bones, contrasts to the plantigrade locomotion of humans and bears, in which the whole foot (toes and metapodial bones) is in contact with the ground.

But not all ungulates live in areas with hard ground – some face the opposite problem, because where they live, the ground is often soft. In winter, Caribou must traverse snow-covered areas, and to help them avoid sinking too deeply, they have the equivalent of snowshoes. Not only are their main hooves exceptionally large, but so are their lateral hooves, which are situated close to the main hooves. Together, these four hooves increase the surface area of each foot, making them effective for travelling over snow and for digging out lichens buried beneath the snow. Arabian and Bactrian camels face a similar problem of moving over soft ground. In these species, both toes on each foot have become enlarged, again increasing the surface area to prevent the camel from sinking too deeply into loose sand.

Selected References: Macdonald 1984, Nowak 2003, Telfer and Kelsall 1984, Vaughan et al. 2011.

A Coat of Many Colours – Camouflage and Communication

An ungulate's coat serves three main functions that we know of: protection from inclement weather, camouflage and communication. To help protect against the elements, the coat consists of two layers of hair – long, coarse outer guard hairs and short, often curly, underfur. The long guard hairs protect the softer underfur by shedding water, a valuable feature in wet areas like coastal rainforests. In many ungulates such as Mule Deer and Caribou, the thick winter guard hairs are also hollow, and the air inside the hair shafts provides extra insulation during the cold winter months. The main function of the underfur is insulation, achieved by trapping air between the short, fine hairs. Through artificial selection, most breeds of Domestic Sheep no longer have guard hairs – only the woolly underfur is left.

In many species, coat pattern provides camouflage and so reduces the risk of predation. White spots or stripes on a brown coat can help the ungulate blend with the shadows and dappled sunlight of the forest floor. The spotted coats of young Elk, Mule Deer and White-tailed Deer are good examples of this type of camouflage (figure 5).

Different coat colours and hair lengths may have communication functions, mainly directed toward members of the same species.

Figure 5. The young in some species of ungulates will hide during the day while their mother feeds elsewhere. (See Appendix 2.) This hiding phase can last up to two weeks.

White-tailed Deer, for instance, raise their tails to expose the long white hairs on the underside and over the rump, then bound off waving the tail from side to side as a warning signal to other deer that a predator is present (colour photograph C-14). Some African antelopes have ears that contrast in colour to the rest of the head; the position of their ears signals aggression or submission, and the contrasting colours help to emphasize ear position. In other species, coat colour or length is related to age. It is not uncommon for young of the year to have quite differently coloured coats from adults (e.g., colour photographs C-6 and C-8). Other coat characteristics may develop only in mature age and sex classes. For example, in some Asiatic wild sheep and goats, only mature males develop large neck ruffs or beards, while in the Indian Blackbuck Antelope, males change from a reddish to a black coat on reaching maturity.

The presence or absence of a contrasting rump patch roughly correlates with the visual density of the habitat relative to the size of the animal and its typical social organization. For example, Elk and Bighorn Sheep have large, pale or white rump patches associated with open habitats (such as grasslands) and group living, whereas White-tailed Deer have small or hidden rump patches because they inhabit visually close habitats and do not normally form groups. The most probable hypothesis suggested for this coat pattern is that light-coloured patches help group members stay together when fleeing from a predator, whereas the absence of a contrasting patch or the ability to hide it will not attract a predator's attention. Larger species that live in open habitats such as Bison on the other hand, probably have sufficient visibility with their dark brown coats that they can maintain cohesion, while their large size limits the number of predator species that regularly prey on them.

Ungulates shed their warm winter coat in early spring and grow a sleeker, shorter one for summer. While shedding, patches of the old winter coat often hang from the body in tattered clumps, giving the false impression that the animal is unhealthy (though some ungulates in British Columbia can suffer from a form of mange that causes hair loss and patches of rough or scab-covered skin). I have seen lactating females shed their winter coat later than other animals in their group, perhaps because of the energy required to produce milk (see colour photograph C-12).

Group Living – Safety in Numbers

Ungulates are social animals, and individuals of most species that live in open areas (such as Stone's Sheep and Rocky Mountain Elk) spend

Figure 6. Most ungulates live in groups, especially species living in open habitats, like these Plains Bison. For individuals, the main benefit of group living is reducing the risk of predation.

the majority of their lives in groups. For an ungulate, the major reason for group living is probably to reduce the chances of being killed by a predator. All the animals in a group can watch for predators, and if one attacks, an animal can hide among the other group members. Because there are always other group members on the lookout for predators, an individual can spend more time feeding than if it were alone and having to be constantly vigilant. If the ungulate is larger than the predator, it will have a better chance to defend itself if it acts together with other group members in cooperative defence. But simple probability – the more group members there are, the lower the chance any particular individual will be killed in an attack – is likely the single most important benefit for ungulates living in a group (figure 6). This dilution effect is probably analogous to our own feelings when hiking in bear country. We feel much safer when with a companion than when alone, probably because it seems our chance of being the one to be attacked is reduced. Even species that live in more closed, forested habitat and do not usually form groups, such as Moose or White-tailed Deer, will form temporary groups in open habitats.

As in most hoofed mammals, adult males and females of all British Columbian species live apart for most of the year, often in different habitats, and come together primarily during the mating season. The most likely reason for this sexual segregation is that the two sexes have

Figure 7. In most species of ungulates, males grow more rapidly and continue to grow for several years longer than females. The adult males of all BC ungulates are larger than the females.

different requirements for food and security. Adult females require safe areas for raising their young, even if this means feeding in areas with less than the best forage conditions. Males, on the other hand, need areas with abundant high quality food resources to maximize their body growth and condition for future competition for mating opportunities. These different requirements are usually met in different habitats and different places, and there is probably the added benefit that separation reduces competition between the sexes.

Different growth strategies of the sexes, called sexual dimorphism, mean males grow larger than females (figure 7). Males seem to be the ones that leave their maternal group, probably voluntarily, whereas young females tend to stay with their mother's group.

Selected References: Bleich et al. 1997, Caro et al. 2004, Corti and Shackleton 2002, Krebs and Davies 1993, Main et al. 1996.

Dietary Habits – What to Eat

Pigs and peccaries are not obligate herbivores – they do not have to eat plants. Although they eat mainly vegetation, they are more omnivorous, and consume everything from roots, bulbs and fruits to bird's eggs and insects. All other ungulates, both Artiodactyl and Perissodactyl, survive almost exclusively on plants, but the types of plants they eat vary. These obligate herbivores are either grazers, browsers or mixed grazer-browsers. Grazers, like Bighorn Sheep and Bison, eat primar-

Figure 8. Grazers, like the Rocky Mountain Bighorn Sheep (left), eat primarily grasses and low growing forbs. Browsers, like the Mule Deer (right), depend heavily on the leaves and twigs of woody shrubs and trees for their food.

ily grasses and forbs; browsers, such as Mule Deer and Moose, eat mostly the leaves and twigs of woody plants; Elk and Mountain Goats are grazer-browsers, eating a mixture of food types (figure 8). Despite their herbivorous diet, some ungulates have been observed eating young ground-nesting birds and other sources of animal protein.

The Challenges of Herbivory – Teeth and Jaws

We can tell much about what an animal eats from its teeth. Ungulates, like humans and other mammals, have heterodont dentition, meaning that they have different types of teeth – incisors, canines, premolars and molars. And like humans, they grow two sets of incisors, canines and premolars during their life, losing their first set (also called deciduous or milk teeth) before maturity. But this point is where the similarities end. Being obligate herbivores, ungulates have highly specialized teeth adapted to gather and process vegetation, and to combat tooth wear. Only the pigs and peccaries, with their more omnivorous diet, have chewing teeth that resemble our own, adapted to masticating a variety of foods.

Compared to other foods, plants are neither very nutritious nor easily digested. These two constraints cause significant problems, especially for large herbivores that must eat lots of vegetation, or for smaller ones that need to eat enough high quality vegetation to meet their high metabolic demands. Both grazing and browsing ungulates have teeth that are modified to deal with their herbivorous diet, and you can easily see these adaptations if you find a skull or lower jaw.

The first problem for the herbivore is to gather or crop enough of the right kind of vegetation to meet its needs. Ungulates have evolved several adaptations to do this. Large grazers and browsers, like Domestic Cattle or Giraffes, wrap their long tongues around a clump of grass or browse, and pull it into their mouths. Others, such as horses and asses, have large upper and lower incisors that they use to crop grasses close to the ground. Small ungulates, like the diminutive Dik-dik antelopes of Africa, have small, slender mouths that allow them to select the most nutritious parts of an individual plant.

All the deer species, as well as the horned Artiodactyls (Bovids), have lost their upper incisors. The upper palate is simply a hard cartilaginous pad. Rather than biting off vegetation like a horse does, these species grasp the plants between the pad and their spade-shaped lower incisors (figure 9), and with a quick upward jerk of the head, snap or tear off a mouthful of vegetation. In some grazing Artiodactyls, especially the larger species, the lower canines have also become modified to resemble the wide cropping incisors, thus increasing the bite size.

Another characteristic of most ungulate jaws is the diastema, the tooth-free gap between the lower canines and the first of the premolars. Its function is uncertain – possibly it helps some browsers, like Moose, strip leaves off branches, but it may simply be a by-product of an elongated jaw that evolved to allow herbivores be more selective feeders. It is even suggested that the elongated face of many ungulates keeps their eyes above vegetation when feeding, so they can watch for predators.

The main problem with eating vegetation, however, is digesting it. The cell walls of plants contain cellulose and often lignin, both compounds that are difficult to digest. It always helps to chew and grind food into tiny pieces, and this is especially true for plant foods. Small food particles have relatively large surface areas, which provide greater

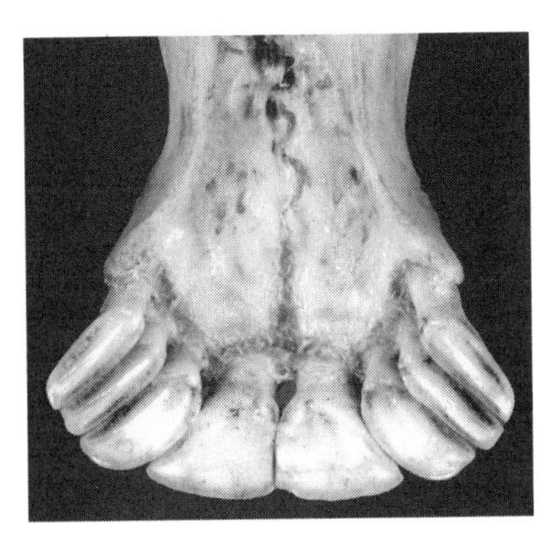

Figure 9. The lower incisors (three on each side) of herbivorous ungulates, such as this Fallow Deer, are wide (spatulate) to help crop forage. The lower canines (the outer tooth on each side) often become incisor-like (incisiform), increasing the size of the deer's bite.

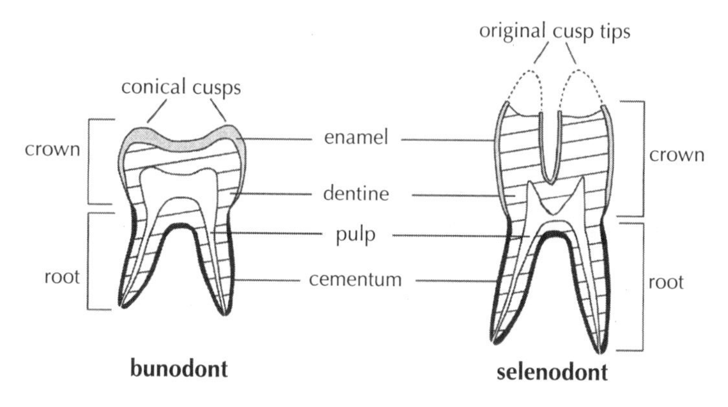

Figure 10. Diagrammatic cross-sections of a human molar (left) showing the typical bunodont crown with low conical cusps, and the molar of a herbivorous Artiodactyl (right) with a selenodont enamel pattern. In the bunodont pattern, the enamel completely covers the crown. In the selenodont pattern, the enamel in the centre of the crown is depressed inward (invaginated). When the cusp tips wear off, hard enamel ridges are left projecting above the softer dentine layers. These ridges are effective for grinding plant food into small particles to aid digestion.

access for enzymes or bacteria to break down the food, thus increasing digestive efficiency. Grinding vegetation into minute pieces requires special teeth, and all herbivorous ungulates share the same basic adaptations of their cheek teeth (premolars and molars).

Enamel is the hardest part of the mammalian tooth, and it is found only on the crowns of teeth. The crown of a pig's or a peccary's tooth consists of low conical cusps covered by a layer of enamel (figure 10). Human cheek teeth have this same bunodont crown pattern. But in a herbivorous ungulate's tooth, the enamel of the crown is highly modified into lateral or vertical folds. The tips of these enamel folds quickly wear off to expose a series of hard cutting edges interspersed with layers of softer dentine. Together, these layers create a ridged millstone, ideal for grinding mouthfuls of plant food into small particles. There are two main enamel patterns that result from these invaginations, crescent-shaped (selenodont) and convoluted (lophodont). These patterns are clearly visible if you look at the top or grinding surface of the cheek teeth (figure 11). The selenodont pattern is found only in herbivorous Artiodactyls, and while lophodont pattern is more typical of Perissodactyls, the premolars of some Artiodactyl have this enamel pattern as well.

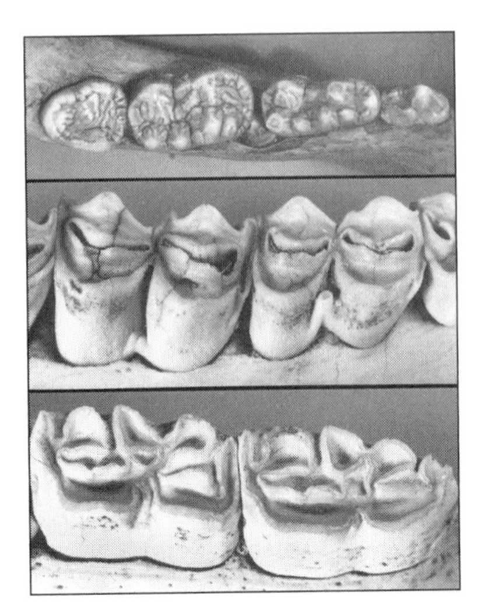

Figure 11. The cheek teeth of many omnivorous and carnivorous mammals have bunodont crowns with low conical cusps and a complete outer covering of enamel, similar to those of pigs (top). In herbivorous ungulates, the enamel is folded; after the tips are worn away, the teeth have a distinct ridges of enamel used to grind plant food. The two main enamel patterns of herbivorous ungulates are the selenodont pattern (middle) found only in Artiodactyls, and the lophodont pattern (bottom) typical of Perissodactyls.

Tooth wear is a major determinant of an animal's life span. The more abrasive the food, the quicker the teeth wear away, and hence the shorter the animal lives. Browsers eat mainly foods such as leaves, buds and young shoots of woody plants that are much softer than tooth enamel, but this is not the case for grazers. The leaves and stems of grasses and forbs growing in dry areas often have a thin layer of fine silica dust on their outer surfaces, and most grasses have silica in their epidermal (outer) cells. Silica, as any woodworker knows, is an excellent abrasive used in sandpaper. Even tooth enamel is not hard enough to resist the constant grinding of silica-laden plants that a herbivorous ungulate must eat. Such an abrasive diet would wear down the teeth quickly and shorten the life span of a grazing animal. Not surprisingly, herbivores have two main adaptations to counter tooth wear. The simplest is that the surface area of the premolars and molars has increased, so that a larger tooth takes longer to wear away. This is well illustrated by the large molars of Elk. But in highly specialized grazers, the crowns of the molars and sometimes premolars, are elongated, while the roots are short (figure 12). Although these high crowned (hypsodont) teeth do not grow continuously like the incisors of rodents, they do continue erupting for several years as the tooth surface wears away. As a result, the hypsodont tooth of a grazer would last longer than the low-crowned or brachyodont tooth of a browser or omnivore eating the same abrasive diet. Because of these adaptations to prolong tooth life, grazers that eat abrasive foods can live as long as similar-sized browsers and omnivores consuming softer foods.

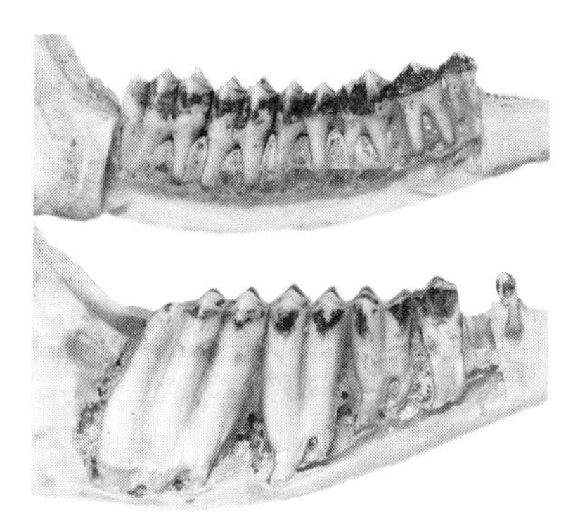

Figure 12. Browsers have low-crowned or brachyodont teeth (top) with root length about equal to the crown height. Grazers have high-crowned or hypsodont cheek teeth (bottom) with very short roots for eating grasses, which are more abrasive.

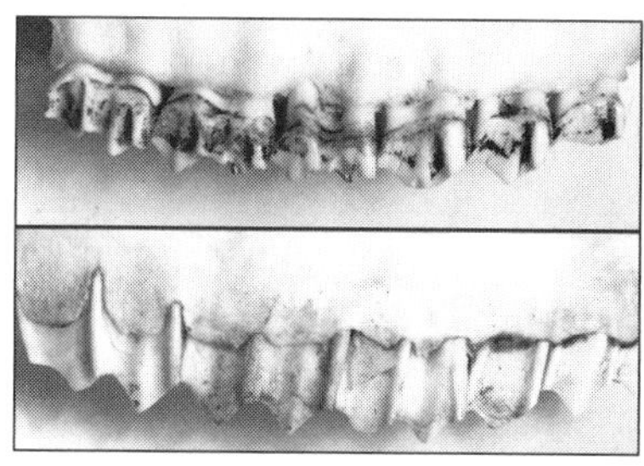

Figure 13. The crown type of an ungulate tooth can be distinguished in the skull by how the tooth enters the jaw. Brachyodont teeth (top) curve inward slightly as they enter the upper jaw, while hypsodont teeth (bottom) continue straight into the jaw.

If you find an ungulate skull or lower jaw when you are out on a hike, you should be able to predict its food habits by examining its cheek teeth, paying particular attention to the molars (the last three teeth). First look down on the grinding surface. Does the enamel-covered crown have low, conical cusps or bumps as in an omnivore like a pig, or are there distinct enamel ridges as in a herbivore? When viewed from above, are the enamel ridges crescent or loop-shaped as in Artiodactyls, or convoluted as in Perissodactyls (figure 11)? Next, look at the side of the molar. Does the crown narrow as it enters the jaw bone, as in the brachyodont tooth of a browser, or does it seem to go straight into the jaw, as in the hypsodont tooth of a grazer (figure 13)? After you've answered these questions you will have determined whether or not it was a herbivore, and if it was, whether it ate mainly browse and forbs, or grasses and forbs.

One final adaptation of ungulates that helps them chew their food, is the hinge or articulating surface (mandibular condyle) of the lower jaw, where it meets the skull. It is flat and wide in herbivores to allow a lateral (circular) movement of the jaw while chewing. This motion optimizes the grinding action of their cheek teeth.

Most mammalian teeth also have features that biologists can use to determine the age of an animal. The first method applies to ungulates under four years old. Like all mammals, ungulates not only grow a set of deciduous teeth, but the timing or succession in which the different teeth erupt is strongly related to the animal's age. Hence age can be estimated from their tooth eruption sequence for the first few years of life. A second method applies to older individuals, and uses one of their permanent teeth. It relies on the fact that the cementum surrounding a tooth's root is deposited at a different rate according to season. In fall and winter, cementum is laid down slowly creating a dense but thin layer, while in spring and summer, the rate is faster and the layer thicker. Like tree rings, these pairs of cementum rings allow a biologist to count the years an animal has lived. A third method of estimating age is by tooth wear; but this method is less accurate, because many factors affect the rate of tooth wear and these can vary among animals, especially with location.

Selected References: Feldhamer et al. 1999, Vaughan et al. 2011, Silvy 2012.

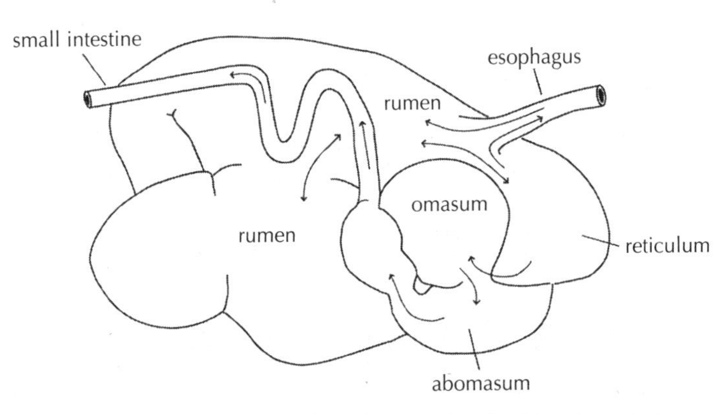

Figure 14. A ruminant's four-chambered stomach (after Vaughan et al. 2011). The abomasum (true stomach) is the last of the four chambers. The rumen, reticulum and omasum are expansions of the esophagus. Arrows indicate how the food passes down the esophagus and enters the rumen where cellulose digestion by micro-organisms and fermentation begins. This involves the breakdown of cellulose into volatile fatty acids that are then absorbed through the rumen wall. As part of this digestion process, food is formed into a bolus (ball) by the reticulum and regurgitated into the mouth to be chewed again (rumination or cud chewing). This grinds the food into fine particles to aid microbial digestion. When the particles are small enough, they pass first into the omasum where water is extracted, and then into the abomasum where protein digestion begins. The partially digested food, along with microorganisms, finally enters the small intestine where protein digestion is completed and nutrients are absorbed.

Extracting Nutrients – Digesting Food

Once food enters the mouth and before digestion can begin, the ungulate's formidable chewing-grinding system breaking it down. Ungulates, like all mammals, lack the enzymes necessary to digest cellulose. Herbivorous ungulates have evolved a way to overcome this shortcoming. They assimilate cellulose and other plant components with the aid of micro-organisms and a fermentation process. The micro-organisms, mainly bacteria, secrete cellulase, an enzyme that breaks down the plant cellulose. This is followed by fermentation, which continues the degradation by creating simpler compounds called volatile fatty acids – the two main ones are acetic acid and propionic acid. The fatty acids are absorbed and transported by the blood system to the liver, where they are metabolized and used as energy sources. But Artiodactyls and Perissodactyls differ in where this microbial fermentation takes place in their digestive systems.

Located just before the true stomach, all Artiodactyls have one or more chambers or false stomachs. Bovids and deer have the most, with three chambers (rumen, reticulum, and omasum) ahead of the abomasum, the true stomach (figure 14), whereas pigs and peccaries have only one small chamber. Bacterial fermentation takes place in the large rumen (fore-stomach), so ungulates using this method of processing food are called fore-gut fermenters. Most Artiodactyls, except for the pigs, peccaries and hippos, are ruminants – so called because they ruminate (figure 15). During rumination, the larger food particles move from the rumen to the reticulum where its honey-combed (reticulated) inner surface forms the food into a ball (bolus). This ball is then regurgitated up the oesophagus into the mouth to be re-chewed, mixed with more saliva, and then swallowed again.

Figure 15. Most Artiodactyl ungulates, like this Bighorn Sheep, are ruminants that regurgitate food to rechew it, so breaking it down into smaller pieces to increase digestive efficiency.

The small food particles are attacked by the micro-organisms, which digest them and break them down into even smaller pieces. When the particles are reduced to a certain size, they pass through a small orifice connecting the reticulum and omasum.

This orifice acts like a sieve and restricts the flow of digesting food. The omasum's folded, muscular walls squeeze most of the water out of the food and resorb it. The resulting drier food mass passes into the abomasum where protein digestion begins, breaking down the food into amino acids. But most protein digestion and amino acid absorption takes place in the intestines. The final step of the digestive process in ruminants involves some additional microbial fermentation in the caecum, located toward the end of the gut.

The ruminant digestive system is almost entirely dependent on micro-organisms for extracting nutrients from the plants. They not only allow it to extract energy from cellulose, but most of the protein digested by a ruminant is actually obtained from the micro-organisms themselves, not directly from the plants. The micro-organisms first use the plant protein for their own reproduction purposes, and then, when they spill over from the rumen into the omasum and beyond, the ruminant begins to digest them. In this way, the fore-gut fermenter gains more valuable amino acids than if it had to rely only on digesting plant proteins.

Perissodactyls, in contrast, are called hind-gut fermenters because all fermentation takes place in their enlarged caecum and colon toward the end of the digestive system. These ungulates chew their food only once and have but a single stomach where digestive enzymes begin to act on the food. The food passes from the stomach to the intestines, where digestion continues to break down proteins into amino acids, and sugars and carbohydrates into glucose; these nutrients are absorbed in the intestines. Any undigested food that reaches the caecum is subject to bacterial fermentation. It is here that the cellulose and other low-digestible plant components are broken down into volatile fatty acids. Like the ruminant, the hind-gut fermenter absorbs the fatty acids and sends them to the liver to be converted into energy.

Fore- and hind-gut fermentation systems result in different benefits and costs. Perissodactyls can increase the rate that food passes through their gut, processing only the most easily digestible part of the food and eliminating (excreting) the less valuable, hard to digest material. This means that, although they must feed almost continuously and are less selective in what they eat, they are able to survive on poorer quality forage than Artiodactyls. But the evolutionary success of modern Artiodactyls over the Perissodactyls is believed to be largely due to the way their digestive system works. Fore-gut fermentation gives Artiodactyls three major advantages. First, the ability to ruminate results in a more effective breakdown of ingested plant material into smaller, more easily digestible particles, which significantly increases the ability to efficiently digest most of the plant matter that they consume. Second, because the microbial fermentation takes place at the beginning of the digestive process, the ruminant can obtain a rich protein source by digesting the rapidly reproducing micro-organisms. And third, the ruminant's large fore-stomach allows it to gather a large amount of food relatively quickly when in the open and more susceptible to predators, and then retire to safer areas to process and digest it. The only disadvantage for Artiodactyls (compared to Perissodactyls) is when food is of poor quality, because rumen fermentation takes longer as food quality declines, and food must be chewed and re-chewed several times until the particles are reduced to a small enough size to allow microbial digestion to occur.

Selected References: Hudson and White 1985, Robbins 1993.

Mineral Licks – Vital Elements

Mineral (salt) licks are important for ungulates. In western North America, licks seem to be particularly high in magnesium and calcium, and to a lesser degree in sodium salts. Ungulates are thought to use licks to maintain mineral balance. This is especially important during spring because growing plants are rich in potassium, a mineral readily absorbed by ruminants. Their bodies respond to this by trying to eliminate excess potassium, but in the process they also excrete magnesium which causes their normally low reserves of this element to drop to critical levels. Acute magnesium deficiency associated with eating lush vegetation is called grass tetany in domestic livestock, and may occur in wild species as well. By using natural licks, wild ruminants may be able to increase their magnesium intake and thus restore their reserves of this vital element. Calcium consumed at the same time is also valuable for milk production in lactating females, bone development in young growing animals and antler growth in deer (figure 16).

In many areas of British Columbia, hoofed mammals use licks in late spring and early summer. Licks can be wet or dry. The presence of a patch of trampled bare earth, sometimes with traces of a white deposit, can indicate that this is a mineral lick. At some dry licks, the substrate is quite hard, and ungulates may chew soft, friable rocks in search of valuable minerals.

Most deer species seem to prefer wet licks, whereas wild sheep and Mountain Goats tend to use dry licks, and Elk use both. Examples of both types of licks can be seen in Kootenay National Park. A wet lick, located on the south side of Highway 93 southeast of Mount Wardle, is used by Moose and other deer, and a little further west

Figure 16. Licking mineral rich soils probably helps ungulates maintain their bodies' reserves of vital elements such as magnesium, calcium and sodium, which appear to be especially important in spring.

toward Kootenay Crossing on the north side of the highway is an extensive dry lick used mainly by Mountain Goats in late spring and early summer.

Selected References: Jones and Hanson 1985, Klaus and Schmidg 1998.

The Social Ungulate: Communication, Morphology and Behaviour

Social organization refers to how animals are distributed in time and space with respect to other members of their species. It encompasses such aspects as group living, mating systems, hierarchies and use of space (e.g., territoriality, home ranges). Unlike its behaviour patterns, which are relatively fixed, a species' social organization is typically quite flexible, responding to environmental factors such as population structure, habitat conditions or resource distribution. Most hoofed mammals live in groups, although a few species may be solitary but not asocial. Like all mammals, they have at least two social periods in their life – when females rear young and when adults of both sexes interact for mating. During these periods, communication plays a key role.

Ungulates communicate with each other in various ways. Earlier, I discussed how morphology, in the form of colour pattern and structure of the pelage, plays a role in communication. Ungulates also use a variety of vocalizations and chemical signals to convey information about their physiological state or social status, and for other forms of social interaction involving competition for mates, courtship during mating and rearing young.

The following sections describe some of the more important social activities of ungulates.

Smells and Signals – Secretory Glands

Ungulates have many glands located in different parts of their bodies (figure 17, table 2). Most are known as epithelial glands because they originate in the epithelium (skin) and are often found as pairs, one on each side of the body. The glands produce odoriferous chemical secretions – pheromones – in the form of volatile chemicals or waxy substances that are used by the animal for a variety of communication functions. The Bison is the only BC ungulate that seems to lack these glands.

Table 2. Epithelial glands in ungulate species in British Columbia. No glands have been reported so far in Bison. (After Chabot 1993 and Muller-Schwartz 1987.)

Species	Pre-orbital	Frontal	Post-cornual	Preputial	Inguinial	Caudal	Anal	Tarsal	Meta-tarsal	Inter-digital
Elk	+					+			+	+
Fallow Deer	+			+					+	+
Moose	+							+	?	+
Mule Deer	+	(+)				+		+	+	+
White-tailed Deer	+	(+)				+		+	+	+
Caribou	+					+		+		+*
Mountain Goat			+							+
Bighorn Sheep	+				+	+	+			+
Thinhorn Sheep	+				+	+	+			+

(+) = poorly developed; ? = conflicting reports; * = hind feet only.

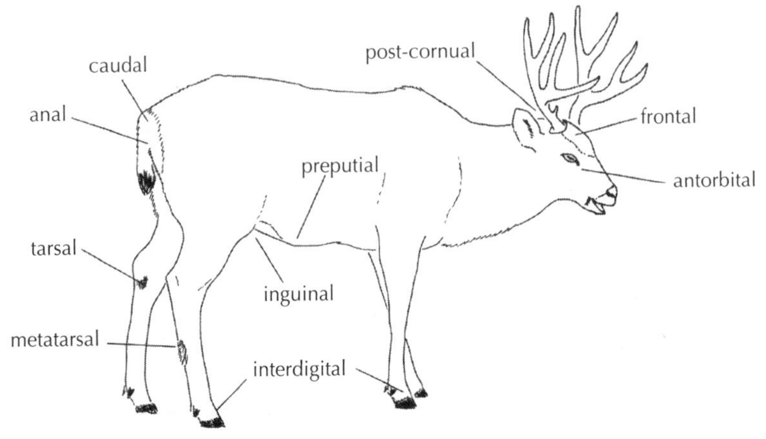

Figure 17. The approximate locations of some common epithelial glands of ungulates. Not all ungulates have all types of glands – the Mule Deer shown here does not have anal, inguinal or post-cornual glands.

Figure 18. Self-marking is common in ungulates. This Rocky Mountain Mule Deer is performing a hock-rub in which he rubs his tarsal glands together while urinating on them, producing a powerful musky odour.

Territorial species often use glandular secretions, along with urine and feces, to mark territorial boundaries. For example, domestic and feral stallions defecate in the same place, creating dung piles that probably serve to define territorial boundaries, or at least advertise their presence as the dominant male. None of BC's wild ungulates are known to hold territories, but some use secretions to mark objects in their environment. Bush-thrashing is a common behaviour, in which the male rubs his horns or antlers vigorously against shrubs or small flexible trees. Scraped bark and broken branches provide visual marks. Some species, such as White-tailed Deer and Mountain Goats, have glands on their heads, and probably leave olfactory signs.

Apart from male mountain sheep, which are thought to use their antorbital glands to mark each other, most BC ungulates seem to use their glands for self-marking or for communicating alarm or physiological condition. The most common types of self-marking are urination and wallowing, which are done mainly by males in the rut (see Species Accounts). Hock-rubbing is another self-marking behaviour, and though commonly practised by both sexes in some deer, is done frequently by males in the rut. When hock-rubbing, the deer hunches his back, hind legs together, then rubs the tarsal glands against each other while urinating over them (figure 18). The urine, coupled with the secretions from the glands, can produce a powerful musky odour that seems to have important social communication functions.

Alarm signals may come from secretions of caudal glands when the tail is raised, or from the tarsal glands when the hairs of the gland are raised. The suggested function of interdigital glands located between the main hooves of many ungulates is to mark trails and bedding sites, and they may also leave odours when males dig their rutting wallows and scrapes. In general, however, the behavioural role of most ungulate glands and their secretions requires further study.

Selected References: Berger 1986, Chabot 1993, Gosling 1985, Müller-Schwartz 1991.

The Battle for Mates – Weapons, Fighting and Displays

All British Columbia's ungulate species are polygamous, or more precisely polygynous. This means that each mating season or rut, a female usually copulates with only one male, but an active male will usually mate with more than one female. The sexes generally follow two different reproductive strategies. Adult females are iteroperous, mating each year as soon as they reach puberty, whereas dominant males are semelparous, mating with several females for only a few years. Less dominant males only occasionally succeed in mating. The mature males that do succeed usually do so after competing fiercely with other males. Such competition among males for females usually involves ritual displays and serious fighting. For animals that we generally associate with eating plants and images of Bambi, hoofed mammals have an impressive array of species-specific weapons.

Charles Darwin was one of the first to recognize that the primary function of an animal's weapons is to use them in aggressive interactions with members of its own species, and that these weapons evolve specifically for this purpose. Only secondarily might they be used in defence against predators. Generally, we think of weapons as having an offensive or attack function – to try to injure or kill an opponent – but many ungulate weapons are also extremely effective for defence. In addition, weapons may also be used for displays in which the animal attempts to intimidate its rival with the size of his weapons rather than fight him.

Although they did not evolve as weapons, hooves can be used in aggressive interactions (colour photograph C-17). Many slender-legged species, such as deer, use their front feet to strike out, sometimes rearing up on their hind legs to deliver flailing blows on an opponent. Some, like Moose or horses, also kick with their hind feet. Hoof strikes can often be seen as small bare patches in an ungulate's coat.

Ungulates have three basic types of weapons that evolved for fighting. We tend to associate long sharp canine teeth with predators (for capturing and killing prey), but several species of hoofed mammals have canines and use them for fighting each other. Pigs and peccaries have sharply pointed upper and lower canines (tusks) that are triangular in cross-section and are directed outward and upward from the jaws. Contrary to popular belief, pigs do not use these tusks to root up bulbs or other buried food – they employ their tough snout for that. Instead, they use their canines to stab at an opponent's head, neck and shoulders. Pigs usually fight head-to-head, and the areas where they receive most of the blows have thick, tough skin to help protect them from the sharp tusks. Other hoofed mammals with canine weapons include horses (mainly in the males), camels (some have a modified premolar on each side of the jaw that acts as an extra canine) and some deer. Although lacking antlers, adult males of both Musk Deer and Chinese Water Deer have long, dagger-like upper canines extending below their lower jaw. Muntjac Deer, on the other hand, have both sharp canines and small antlers. Each of these deer species use their canines for fighting, but others, such as Elk or Caribou, have only vestigial (non-functional) canines that have no value as weapons.

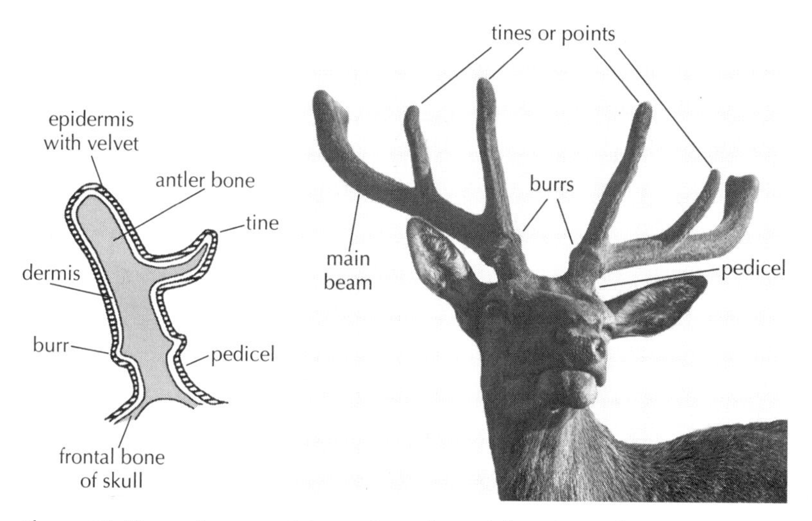

Figure 19. The main parts of the antlers of an adult male Rocky Mountain Elk. A hair-covered skin called velvet covers the antlers while they grow; it contains blood vessels and nervous tissue to feed the growing antlers.

Antlers, the second ungulate weapon system, are unique to modern deer, though not all species have them. The general antler plan consists of a ring at the base, called a burr, with a main branch or beam, round or oval in cross-section, extending upward from it. Depending on the species, the main beam either has series of branches (tines or points) along its length, as in Elk, or becomes palmated (flat and wide), as in Moose and Fallow Deer (figure 19). These paired, bony structures grow from projections called pedicels, located on the frontal bones of the skull. During growth, the antler develops under a layer of specialized skin. This antler skin is called velvet, because of its covering of short, soft hairs (colour photographs C-2 and C-3). It contains many sebaceous and scent glands, and is richly innervated so the antlers are highly sensitive when growing (C-2). Although there are vessels within the bone tissue of growing antlers, most of the nutrients are carried by the blood vessels lying on the surface just beneath the velvet.

Antlers grow faster than most other tissues and can increase in length by up to 20 mm per day. Most mineralization (hard bone development) occurs toward the end of the growth period when levels of the male hormone testosterone increase in the blood. Shortly after mineralization is complete and the antler bone is dense, the velvet dies and separates from the antlers. The deer then rub off the dead velvet

Figure 20. Antlers are shed each year before new ones can be grown. This yearling male Rocky Mountain Elk has just lost one of his antlers.

against trees or shrubs, leaving the antlers with a hard, clean surface (colour photograph C-4). At this stage, the antlers are dead bone with no nervous tissue and only a residual, inner blood supply that is soon cut off.

After the mating season, testosterone levels circulating in the body decline. A thin layer of bone at the junction of the pedicel and the burr loses minerals and weakens, and eventually the antler breaks off leaving the pedicel behind (figure 20). The timing of antler loss depends on the species and on the age and condition of the animal, and while their shape seems primarily controlled by genetics, their size is affected by nutrition. Antlers also increase in size and complexity as the animal gets older, but the number of points cannot be used to estimate the age of an individual, except for a yearling, which may have only a single spike and no branches.

Except for Caribou, only male deer grow antlers, though very rarely a female may develop a small pair. Female Caribou are unique because they grow antlers, and they keep them longer than males. Biologists are not sure why female Caribou grow antlers, but it is likely that they give them an advantage over males when competing for food in late winter and early spring.

In most species, growing and shedding antlers each year represents a tremendous investment of nutrients, especially for adult males. The antlers of species like Elk and Moose can account for more than five per cent of their total body weight. Biologists believe that shedding and regrowing new antlers each year provides two important advantages over a permanent set. First, when antlers have lost their skin cover and stopped growing, they cannot begin growing again. If pieces were to break off, as they often do in fights, permanent antlers could become smaller as a male got older. Second, the first set of antlers grown by a male would also be his last, and though his body would continue to grow, his antlers would not. By shedding and regrowing antlers each year, a male not only grows a new set of unbroken antlers, but he also grows larger ones to keep pace with his body growth. In very old males, though, antler size may decrease.

Antlers are one of the best examples of the dual offence-defence functions seen in most ungulate weapon systems. Many sharp-pointed tines make antlers a formidable weapon, capable of inflicting piercing blows. But the branching structure of antlers also provides a good defence against an opponent's weapons – the antler forks can catch against the opponent's forks.

Generally, male deer fight head-to-head, pushing and twisting as they try to throw each other off balance, so they can then thrust their antler points into their opponent's side or rump. But ungulates only use their antlers for fighting when they are hard. Growing antlers are easily damaged; so, during this period, males fight with their front feet.

In some countries, including Canada, deer are farmed not only for meat (venison) but also for their antlers. Both velvet and the antler bone are believed to have medicinal value, although velvet usually commands a higher price. Velvet is also reputed to have an aphrodisiac effect, the active ingredient being pantin or pantocrin.

The third group of ungulate weapons are horns. Four families of hoofed mammals have horns, and each has its own unique type: the Rhinoceroses (Rhinocerotidae), the Giraffes (Giraffidae), the Bovids (Bovidae – cattle, sheep, antelopes, etc.) and the American Pronghorn (Antilocapridae). Rhinoceroses are the only living Perissodactyls with horns, which consist of a solid mass of keratinized epidermal cells resembling thick hairs cemented together by other epidermal cells. These horns grow from roughened areas on the nasal and frontal bones of the skull. Rhino horns are never shed but will slowly re-grow if damaged or cut off.

The horns of the Giraffe and Okapi are unbranched, bony, skin-covered protrusions that extend from the skull at the junction of the parietal and frontal bones. Both sexes have them. Giraffe horns are larger, consisting of a pair on the top of the head between the ears; mature males usually have a third horn on the forehead between the eyes. Giraffes use their heads as clubs, swinging their long necks and trying to hit an opponent with these bony bumps.

A third horn type is most characteristic of Bovids; in fact, *bovid* means hollow-horned. These horns have two main parts: an inner bone core attached to the frontal bones of the skull, and an outer keratin horn sheath (figure 21). In addition, between the horn sheath and the bone core are two thin layers of tissue, the epidermis and the dermis. Like antlers, Bovid horns are paired structures that grow each year, but unlike antlers, they are neither branched nor shed. Instead, the new year's horn sheath of keratinized epidermis grows over the bone core and under the previous year's sheath. The result resembles a series of ice-cream cones stacked one inside the other. Except for the first sheath, what we see on the outside is only the basal part of each annual horn sheath.

Horn growth usually ceases in late fall or early winter and begins again in spring. In some species this creates a distinct ring (annulus) mark, where one sheath stopped growing and the next began. By counting these annual rings, we can estimate the age of individuals in species like wild sheep and Mountain Goats. But the first ring in some individuals can be difficult to recognize, and estimating the age of females beyond about five years is difficult because the rings are very close together and hard to distinguish from other minor rings and bumps on the horn's surface.

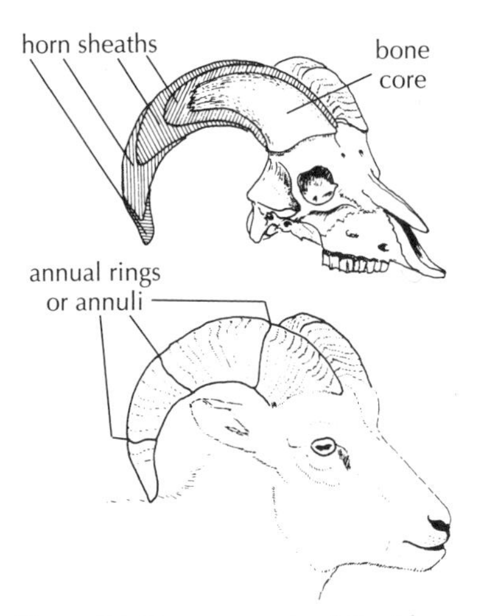

Figure 21. The two parts of Bovid horns: a bone core attached to the skull and a horn sheath that grows over the core each year. The breaks or ridges between annual horn sheaths form rings that can be used to determine the age of an animal. Counting annual rings is usually easier on the horns of males, and it is a better measure of age in some species than in others.

Most male and female Bovids grow a pair of horns (although not the aptly named Four-horned Antelope of India that grows two pairs), and in most species the adult male's are larger. The Bovids have three basic fighting styles, depending on their horn types, and it is reasonably easy to predict how a species fights from its horn structure. Horns that are short, smooth and sharp like those of Mountain Goats (figure 22) are potentially dangerous weapons capable of inflicting lethal wounds. Animals with these horns use them when standing side by side, usually head to tail, sometimes stabbing at the opponent's flanks. But with short, dangerous weapons like these, the animals try to avoid physical contact as much as possible. Instead, they use threats, trying to convince each other to back down and leave without a fight.

Few Bovids have short, dangerous horns. More common are the wrestlers, whose horns are longer, often twisted in shape and covered with many bumps and ridges. There are no BC species with this horn type, but African antelopes such as the Greater Kudu and Common Eland are good examples. The twisted shape and surface ridges act like the forks of a deer's antlers, and help catch and hold an opponent's horns. Like deer, these Bovids fight by wrestling head-to-head, each animal trying to twist the other off balance and then stab him with his horn points.

Figure 22. The Mountain Goat is one of a few ungulate species that have short, sharp horns.

The third type are the bashers, easily recognized by their massive horns. Typical bashers are male Bighorn and Thinhorn sheep in BC, the Muskox of the Arctic, and the Alpine Ibex of Europe. Combatants use their horns as clubs to smash head-to-head until one of them concedes. Animals that fight like this often increase the force of the blow by running toward each other, standing on the hind legs before dropping into the clash, or even trying to get the advantage of the higher ground (figure 23). Some species also have internal skull adaptations that help absorb the tremendous impacts (see the Bighorn Sheep species account).

The American Pronghorn, a unique North American ungulate often incorrectly called an antelope, has horns similar in structure to those of Bovids. This species is the last surviving member of a group that flourished in North America in the Miocene and early Pliocene. While the horns of the living Pronghorn also have a horn sheath growing on a bone core, they differ in two ways from those of Bovids. First, unlike the unbranched horns of Bovids, the horns of the male Pronghorn are forked – hence the animal's name. Second, Pronghorns shed their horn sheaths each year. The new sheath begins growing in October after the rut, while the previous sheath is still attached. The old sheath is then shed a few weeks later in November.

The defensive functions of most ungulate weapons mean that serious injury is the exception rather than the rule. Males try their best to hurt each other, especially when competing for females, but they usually fight weapon-to-weapon, trying to defend themselves while they attack – it is difficult to injure an equally matched opponent when fighting with such dual-purpose weapons. When unequal males meet, they invariably avoid fighting each other, because the smaller or younger male concedes to the larger or older one.

Despite the defensive value of weapons, fighting can be expensive for ungulates. Even if an animal is not killed or injured, it takes a lot of energy to fight; energy that might be better spent looking for mates or for food. Other than simply walking or running away, animals will use other tactics to avoid fighting. They often use displays to convince an opponent to give up and leave, or at least to abandon an attack. Displays are behaviour patterns that are conspicuous and usually oriented toward another individual. Some are even derived from other behaviours that originally had a different function. Displays can be purely behavioural (postures, vocalizations, etc.) or involve specialized morphological features that have evolved as part of the display. Three broad classes are associated with fighting in ungulates: threat, dominance and submissive displays.

Figure 23. A male Bighorn Sheep performs a threat jump, a fighting behaviour used before head-to-head clashes. He has moved uphill to use gravity to help increase the force of the blow.

Threat displays are usually aggressive movements – signalling that the animal is ready to fight or to defend itself. Invariably, threats involve the animal's weapons, its fighting style or both. A man shaking his fist is a threat – he is displaying his weapon (his fist) and by shaking it is showing how he will use it. In ungulates, the same principles hold. A male Bighorn Sheep will raise its head and point the top of the horns toward the opponent, signalling he is ready to use them, or even rise up on his hind legs as a prelude to a head-to-head clash (figure 23). Similarly, most deer (including males without hard antlers) raise their chin with neck extended and ears back, as a low level threat. This can escalate: first, the animal rears up on its hind legs; if this threat is ignored, it may strike out at the opponent with its forefeet. Ungulates also use noises and vocalizations as threats. Male deer will sometimes

Figure 24. Not all threats are at first obvious to human observers. A Bison raises its tail as an aggressive signal.

thrash their antlers against a bush or small tree to make a noise threat. Species that bite, such as some deer and pigs, will grind their teeth or snap their jaws. Many ungulates use vocalizations that may or may not also accompany a physical threat. These range from the roar of a male Bison or the bugle call of an Elk to the rush-snort of a Mule Deer.

In dominance displays, the animal tries to look as large as possible. Because ungulates look largest when viewed from the side, many dominance displays are performed when the actor is standing at right angles to an opponent – these are called lateral or broadside displays. They are often accompanied by a slow, stiff-legged walk that probably draws attention to the display. Evolution can favour morphological adaptations that enhance lateral displays. Some ungulates, such as Moose or Rocky Mountain Mule Deer, can raise the hair along the top of the neck and back to make them appear bigger and more threatening. In others, such as male Bison, the spines of the thoracic vertebrae are especially elongated, increasing their shoulder height and, thus, their lateral profile. Some displays used in aggressive situations are not obviously (at least not to humans) related to fighting or dominance, such as the raised tail of a Bison (figure 24), or the pawing and wallowing of both male Domestic Cattle and Bison. Besides their value as weapons, antlers and horns may also serve a display function – their size being an indication of the bearer's size, age, physical condition and probable social status.

Figure 25. Many submissive displays are the antithesis of aggressive displays. Here, a female Rocky Mountain Elk lowers her head as a dominant male approaches. She is also making biting movements (see page 100).

The goal of much submissive behaviour is probably to reduce aggression in an opponent. Ungulate submissive displays usually entail one of three strategies: mimicry, appeasement and antithesis. Submission based on mimicry (e.g., of offspring behaviour) is rare in ungulates. Bovids and Equids (horses and zebras) use appeasement, where the subordinate animal licks, nuzzles or rubs against the dominant one. Probably the most frequent submissive displays of ungulates are the antithesis (opposite) of threats. When performing these displays, the subordinate animal acts as unthreatening as possible. It may lower its head or even crouch, making itself appear small, or hide its weapons. Most species of ungulates in British Columbia use a low stretch or head-low pattern (figure 25) to signal submission. Males also use this non-threatening pattern during courtship to approach a female.

Selected References: Bubenik and Bubenik 1990, Darwin 1899, Geist 1966, Geist 1986, Goss 1983, Krebs and Davies 1993, O'Gara and Matson 1975, Walther 1984.

Reproduction and Mating Systems

Wild ungulates in British Columbia breed only once a year. The main determining factor for when mating takes place is the birth period. The newborn has the best chance for survival if born when annual plant growth is just beginning and most nutritious, which is early spring in most parts of BC. Being large mammals, ungulates have a long gestation period, and for spring births, mating must usually take place in the fall or early winter. This is also when ungulates are usually in their best physical condition, just after the end of summer.

Ungulates have a variety of mating systems, and in all species it is the males who are most active. This is because of the different investments the sexes make in offspring. Only females rear the young, and so they must invest heavily in their offspring in the form of milk and protection. As a result, a female needs to choose the best – genetically fittest – male to mate with. Male ungulates, with little direct investment in young other than genes, do not need to be choosy; instead, it pays for them to mate with as many females as possible. This is why they compete with each other, often intensely, for the opportunity to mate. It can also lead to the evolution of elaborate weapons and displays, and to the evolution of different mating systems, which are often responsive to environmental rather than genetic factors.

In the mating season, males of some ungulate species defend territories that attract females because they contain resources such as food or safety from predators. Males of other species congregate in large numbers and hold very small territories, so females might compare potential mates more easily. In BC, male ungulates follow one of two mating systems: they either defend and court a single female (a tending pair), or they defend a harem of females, mating with each as they come into heat (estrus).

Selected References: Krebs and Davies 1993, Walther 1984.

Courtship – Persuasive Partners

There is more to the mating period (rut) than the simple matter of males and females coming together to copulate. Each sex has certain challenges to overcome. A male ungulate must find a female coming into estrus before she has already mated, while a female needs to choose a suitable mate from numerous eager suitors. This involves courtship. Professor Niko Tinbergen, one of the founders of modern ethology (animal behaviour), proposed that courtship can have four main functions: orientation, persuasion, synchronization and reproductive isolation. For ungulates, orientation probably has a limited role, whereas persuasion is probably the most evident.

The courtship behaviour of male ungulates is usually obvious, in stark contrast to the seeming passivity of the females. Some biologists believe that evolution has selected behaviours in both sexes that serve to manipulate the other partner during courtship. While selection appears to favour male courtship behaviour that attempts to persuade females to copulate with them, most research suggests that females actually determine whether most copulations and inseminations are successful. Males have evolved to perform behaviours that reflect their biological fitness. Indicators of fitness can be elaborate and prolonged courtship, the ability to defend a territory with valuable resources such as food or security from predators, or being able to defend a female against other males. In contrast, females in most ungulate species seem to have evolved to be reluctant partners, which, in effect, manipulates the male to provide evidence of his fitness. She accepts copulation only from a male who courts her for a long time, provides her with resources or successfully defends her from other males. Females rarely need to perform courtship, because males are eager to mate with as many females as they can in a mating season; nevertheless, females have sometimes been observed seeking out males. Males usually compete intensely for females. Fights during the peak mating season are often especially vicious and dangerous, more so than before the rut, because a courting male must guard his mate from competitors until fertilization is assured.

A male ungulate uses chemical cues in the urine to detect a female coming into early heat. With submissive or non-aggressive behaviour patterns, he approaches the female and sniffs her ano-genital region. Invariably, the female responds by urinating. The male then sniffs the urine and performs a lip curl (flehmen behaviour), standing for several seconds with his head raised and mouth open (figure 26). We are reasonably certain that by performing this behaviour, the male is using a pair of special glands called vomeronasal or Jacobson's organs

Figure 26. A male Bighorn Sheep uses a lip curl to detect chemicals in a female's urine for signs that she is coming into heat.

Figure 27. In hoofed mammals, the males perform most of the courtship behaviours. Here, a male Rocky Mountain Elk approaches a female, flicking his tongue and uttering soft vocalizations (see page 100).

located in the roof of his mouth. These glands are especially sensitive to chemicals in the female's urine. While the male is lip curling, a female not coming into estrus usually takes this opportunity to move away. But if she is in heat, the male will follow and court her.

The courtship patterns (figure 27) of male ungulates can be few and simple, or numerous and varied, depending on the species and its mating system. Those that defend harems of two or more females, or hold mating territories, tend to have briefer and less complex courtship patterns than species that form tending pairs. In tending pairs, and to a lesser extent in other mating types, courtship involves a series of behaviour patterns that increase the physical contact between the sexes and culminate in copulation. The various courtship patterns and mating systems used by BC ungulates are described in the Species Accounts.

Selected References: Geist 1971, Krebs and Davies 1993, Tinbergen 1953, Walther 1984.

Rearing Offspring – Maternal Care

Being mammals, ungulates rear their young on milk. Because only females produce milk, they are solely responsible for rearing offspring (colour photograph C-8). Males in most ungulate species contribute nothing beyond genes to their young. Hoofed mammals in British Columbia have gestation periods ranging from 147 to 178 days in Mountain Goat to 290 days in Bison. At the end of the long pregnancy, young ungulates are born precocial – meaning that their eyes are open and they can walk and run shortly after birth. Their birth weights range from as low as 2 kg in Mule and White-tailed Deer twins to 18 kg for the single young of Bison.

During the first few weeks of life, ungulate young can be divided into two types: followers and hiders. Young followers, such as Bison, simply stay close to their mothers throughout the day. Young hiders, such as Mule Deer, usually have coat coloration patterns that help camouflage them. The mothers of hiders lead them to the general area where they will leave them, and then the young selects the hiding spot (colour photograph C-7). It may change hiding places during the day, so on returning to the general area, the mother must call to her young to find it. She may return briefly once or twice during daylight to nurse her young. So, except during the night, she spends little time with her offspring beyond that required for nursing and cleaning. Only later when the young is larger and needs solid food, does it accompany its mother. It is this hiding behaviour that many people misinterpret as abandonment when they find a young deer fawn lying in the bush. In almost every case, the young has not lost its mother, she has simply left it for a few hours and will return – so the youngster should be left alone and not disturbed. (Appendix 3: Summary of Birth Data shows which strategy each of BC's hoofed mammals uses.)

Selected References: Krebs and Davies 1993, Walther 1984.

Hoofed Mammals in British Columbia

British Columbia has the greatest biodiversity of any province in Canada. This variation in natural systems results primarily because the province lies at the eastern edge of the North Pacific Ocean and is dominated by mountain systems, most of which run north-south (figure 28). British Columbia covers 950,000 square kilometres and extends over 11 degrees of latitude. Its size, shape, location and topography are major influences on the province's climate. The climate is wettest along the coast, because moist air masses moving eastward from the Pacific Ocean cool as they rise and pass over the Coast Mountains. As a result of this cooling, moisture in the form of rain and snow falls mainly on the western slopes of this mountain range. This in turn creates a relatively dry rain shadow along the eastern

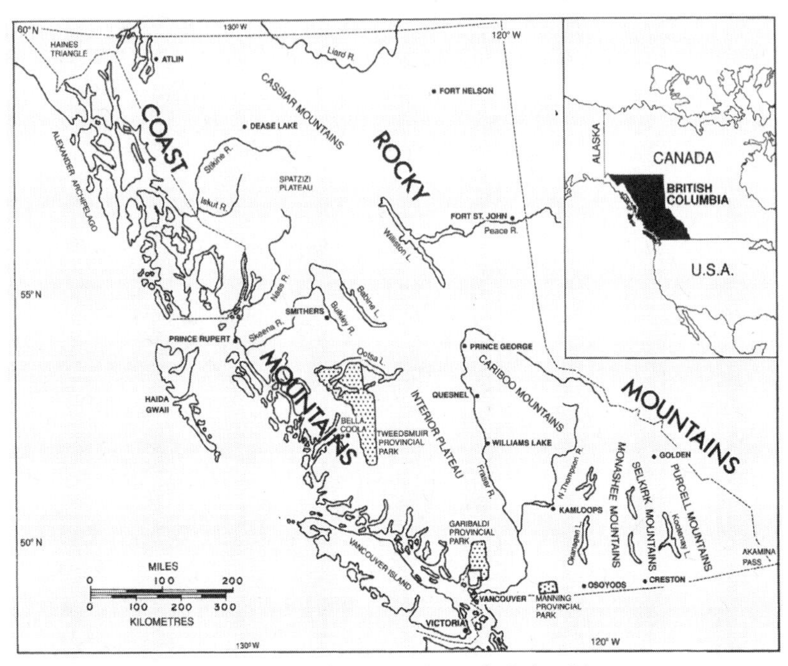

Figure 28. General geographic features of British Columbia.

slopes of the Coast Range, and as these drier air masses descend, they become warmer and take up moisture from the land below. This effect helps create the relatively arid conditions of the interior plateau of BC, especially in areas like the Okanagan and Thompson River valleys. This pattern of high precipitation on western slopes, and drier eastern slopes and intervening valleys, tends to be repeated across the province's many mountain ranges as far east as the Rocky Mountains on the border with Alberta. A good example of the rain-shadow effect can be seen when driving east along Highway 93 through Manning Provincial Park. When the highway begins to descend the eastern side of the Coast Mountains toward Princeton, the wet, dense coastal forests of the western slopes give way to dry, more open Douglas-fir and Ponderosa Pine forests.

Although BC enjoys a high level of diversity in habitat and in plant and animal species, its rich flora and fauna live somewhat precariously. This is because about 75 per cent of the province is more than 1,000 metres above sea level (figure 29). Average temperature drops with increasing altitude, which in turn significantly affects the life cycles of plants. Consequently, much of the province has relatively

Figure 29. The Coast Mountains looking west from an area northwest of Lillooet.

Figure 30. Grasslands near the confluence of the Fraser and Chilcotin rivers southwest of Williams Lake.

low biological productivity. The most productive lands are below this elevation and are important not only for wildlife, but also for humans. In these lowlands we conduct most of our forestry, agricultural and mining activities, and build most of our dams, reservoirs and settlements. All these activities create tremendous competition for the province's limited productive land. Most of BC's ungulates spend the winter in low elevation habitats, so the maintenance of winter ranges is critical for their survival. In this province, grasslands are one of the most threatened ecosystems important for ungulates (figure 30).

To help understand BC's rich biodiversity, provincial biologists use two ecological classification systems. The first, originally developed by Vladimir Krajina and used by the BC Ministry of Forests, is comprised of 14 biogeoclimatic zones (the colour map can be found on the Ministry of Forest website, http://www.for.gov.bc.ca/hfd/ library/ documents/treebook/biogeo/biogeo.htm or by typing "biogeoclimatic

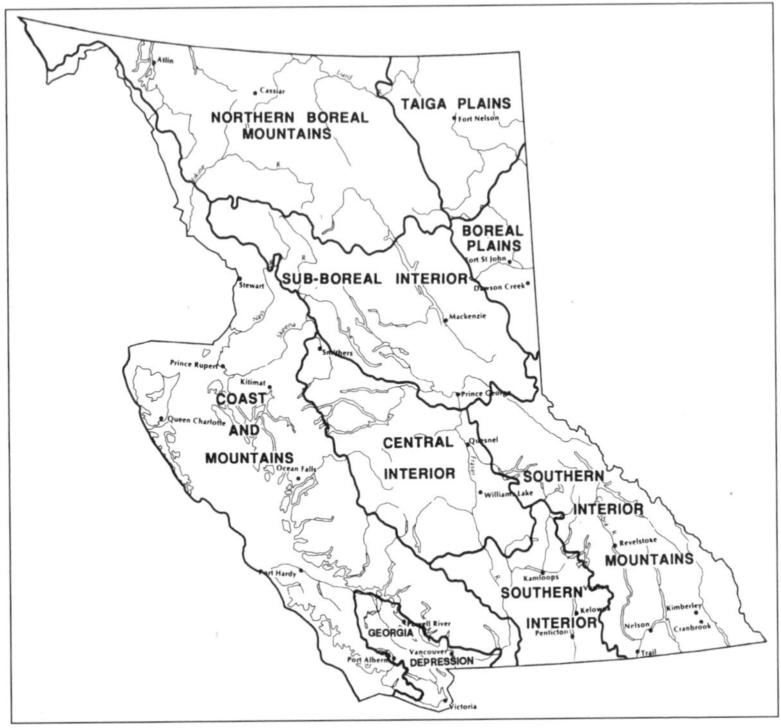

Figure 31. The nine terrestrial ecoprovinces of British Columbia. (Ministry of Environment, 1995.)

zones of British Columbia" into your search engine), delineated by their relatively similar climates and vegetation. In mountain regions of the province, these zones are strongly related to elevation. The second, developed by Dennis Demarchi and adopted in 1985, is the BC Ministry of Environment's system of ecoprovinces (figure 31). This scheme consists of 10 broad-scale geographic areas, each with a relatively consistent climate and topography. Each ecoprovince is subdivided into ecoregions and ecosections.

Wild ungulates are found in all of British Columbia's 9 terrestrial ecoprovinces (table 3) and 14 biogeoclimatic zones (table 4). Of the ecoprovinces, the Southern Interior Mountains has the greatest diversity of ungulates, with 8 subspecies making their home there. Of the biogeoclimatic zones, the Alpine Tundra Zone is used by the greatest variety of ungulates – 14 subspecies – although it is not a major zone for all of them, and the Coastal Douglas-fir Zone by the least – 4 subspecies. Alpine habitats are important to many ungulates

Table 3. Distribution of 18 subspecies of hoofed mammals in the nine ecoprovinces of British Columbia.

ECOPROVINCES	Coast and Mountains	Georgia Depression	Southern Interior	Central Interior	Southern Interior Mtns	Sub-Boreal Interior	Northern Boreal Mtns	Boreal Plains	Taiga Plains
Rocky Mountain Elk *Cervus elaphus nelsoni*	(p)*		P	(p)	P	(p)	(p)	P	P
Roosevelt Elk *C. e. roosevelti*	P	P							
Fallow Deer *Dama dama*		P*							
Northwestern Moose *Alces alces andersoni*	P		P	P	P	P	P	P	P
Alaskan Moose *A. a. gigas*	(p)						P		
Yellowstone Moose *A. a. shirasi*					P				
Columbian Black-tailed Deer *Odocoileus hemionus columbianus*	P	P							
Rocky Mountain Mule Deer *O. h. hemionus*	(p)		P	P	P	P	(p)	P	P
Sitka Black-tailed Deer *O. h. sitkensis*	P								
Dakota White-tailed Deer *O. virginianus dacotensis*						(p)	(p)	P	P
Northwest White-tailed Deer *O. v. ochrourus*		(p)*	P	(p)	P	P			
Woodland Caribou *Rangifer tarandus caribou*	(p)			P	P	P	P	P	P
Wood Bison *Bison bison athabascae*								P	P
Plains Bison *B. b. bison*							P*	P*	
Mountain Goat *Oreamnos americanus*	P	P	P		P	P	P	(p)	(p)
Rocky Mountain Bighorn Sheep *Ovis canadensis canadensis*			P	P	P				
Dall's Sheep *O. dalli dalli*							P		
Stone's Sheep *O. d. stonei*						P	P		P

P = present; (p) = minor presence; * = introduced.

Table 4. Distribution of 18 subspecies of hoofed mammals in the 14 biogeoclimatic zones of British Columbia.

ZONES	Alpine Tundra	Bunchgrass	Boreal White & Black Spruce	Coastal Douglas-fir	Coastal Western Hemlock	Engelmann Spruce - Subalpine Fir	Interior Cedar-Hemlock	Interior Douglas-fir	Mountain Hemlock	Montane Spruce	Ponderosa Pine	Spruce-Willow-Birch	Sub-Boreal Pine-Spruce	Sub-Boreal Spruce
Rocky Mountain Elk *Cervus elaphus nelsoni*	P	(p)	P			P	P	P		(p)	(p)	P	(p)	(p)
Roosevelt Elk *C. e. roosevelti*	(p)			(p)	P				P					
Fallow Deer *Dama dama*				P*										
Northwestern Moose *Alces alces andersoni*	P	(p)	P		P	P	P	P	(p)	P	(p)	P	P	P
Alaskan Moose *A. a. gigas*	P		P	(p)	(p)				(p)			P		(p)
Shiras' Moose *A. a. shirasi*	P					P	(p)	P		P	P			
Columbian Black-tailed Deer *Odocoileus hemionus columbianus*	P			P	P				P					
Rocky Mountain Mule Deer *O. h. hemionus*	P	P	P		(p)	P	P	P	(p)	P	P	P	P	P
Sitka Black-tailed Deer *O. h. sitkensis*	(p)				P				P					
Dakota White-tailed Deer *O. virginianus dacotensis*			P			(p)			P					
Northwest White-tailed Deer *O. v. ochrourus*	(p)	(p)				P	P	P		P	P			P
Woodland Caribou *Rangifer tarandus caribou*	P		P			P	P			P		P	P	P
Wood Bison *Bison bison athabascae*			P									P		
Plains Bison *B. b. bison*			P*									P*		
Mountain Goat *Oreamnos americanus*	P		(p)		P	P	P		P	(p)	(p)	(p)		(p)
Rocky Mountain Bighorn Sheep *Ovis canadensis canadensis*	P	P				P	(p)	P		(p)	P			
Dall's Sheep *O. dalli dalli*	P											P		
Stone's Sheep *O. d. stonei*	P		(p)			P						P		

P = present; (p) = minor presence; * = introduced.

that spend their summer at high elevations in pursuit of good feeding conditions for laying down enough fat to survive winter. Although the Coastal Douglas-fir zone has the least number of ungulate species, it is a productive habitat. It covers a narrow strip along the southeastern coast of Vancouver Island and a very limited area on the islands in the Gulf of Georgia and southwest coast. Although no subspecies is found in every biogeoclimatic zone, Northwestern Moose and Rocky Mountain Mule Deer live in all but one, and Rocky Mountain Elk and Mountain Goat are each found in 10 or more. This range of habitats reflects the general adaptability of hoofed mammals to widely varying environmental conditions.

Selected References: Cannings and Cannings 1996.

Exotic Ungulates in British Columbia

Two categories of exotic (non-native) ungulates occur in British Columbia – feral and introduced species. A feral species is a domestic species now living free without human control. There are three species of feral ungulates in the province: horses, sheep and goats.

Feral horses are found mainly in the central interior, west of Williams Lake. Some were present in the interior when European explorers first arrived in BC. Presumably, they originated from horses brought to the southwestern United States by the Spanish in the 17th century; later, First Peoples brought these horses northward. Feral horses are considered pests by some people, so they are hunted and rounded up.

Domestic Sheep and Goats were common livestock raised along BC's Inside Passage. Most feral populations of these species probably resulted from animals that escaped or were abandoned when farms closed. A small number of feral sheep still live on Saturna, Saltspring and Lasqueti islands. Feral goats are found on Jedediah, Lasqueti, Saturna and Texada islands, and may still exist on Pender Island. A feral goat population on Saltspring Island appears to have been eradicated in the late 1980s, and one on Sidney Island has also been exterminated.

Feral cattle were once found in various parts of the province including the Tofino area, Estevan Point and Cape Scott on the west coast of Vancouver Island, and on the northern and eastern parts of Graham Island in Haida Gwaii. It's almost certain that feral cattle no longer exist on Vancouver Island, although there are unconfirmed reports of some on an island off the west coast. Only a small population of about 12 animals possibly still survives in the eastern part of Naikoon Provincial Park, slightly north of Tlell on Graham Island. The first cattle on Haida Gwaii were released in 1886 by the Hudson's Bay Company near Massett. Then at the beginning of World War I, the feral population increased when settlers leaving for the war freed their animals. More recently, at least two herds were abandoned as late as the 1960s. These releases explain why there are traces of most major British dairy and beef breeds among the remaining feral animals. But long ago, they lost all traces of tameness and are harder to approach than the local deer.

The most unlikely feral species to have occurred in BC was the Bactrian Camel, the large shaggy-haired, two-humped camel of central Asia. During the Cariboo Gold Rush, around 1862, 22 camels were imported as pack animals. They proved unsuitable and some were released in the Lac la Hache and Westwold areas. The small feral population appears to have died out around 1905.

One domestic ungulate species that almost became feral in the province was the Wild Boar. Some of these were illegally introduced into an area near Harrison Hot Springs and others on the east side of Cultus Lake. Fortunately, the released animals did not survive –if they had, they could have become a serious pest problem. Introductions of non-native species are almost always a biological and economic disaster, and so must be avoided at all costs. It is important to recognize that as long as there is at least one of each sex in a small group, it can grow to a large population and become widespread. A recent analysis by D.M. Forsyth and R.P. Duncan of ungulates introduced into New Zealand has shown that as few as a single pair of ungulates can result in a successful introduction, and that liberations of six or more animals were always successful. With so few founder animals capable of establishing a large population, the eradication of introduced ungulates can be very difficult, if not impossible, and certainly economically costly. These recent findings underscore the importance of avoiding introductions (whether planned or accidental) of non-native species.

Two non-native wild species have also been introduced and allowed to roam free in parts of BC, but only one still survives. European Red Deer brought from New Zealand in 1914 and in 1918 were released near Massett on Haida Gwaii. The population increased until the beginning of the Second World War, but then declined, presumably from overhunting by armed forces personnel. There are no recent records of any survivors. The second introduced species is the Fallow

Deer, and populations still exist on some of the southern Gulf Islands along with several substantiated reports in the lower Fraser River valley (see the European Fallow Deer species account, page 110).

Selected References: Cannings and Cannings 1996, Carl and Guiguet 1972, Cowan and Guiguet 1965, Shank 1972.

Conservation and Wildlife Management

Wild ungulates have always been an important natural resource in British Columbia. Moose, deer and Elk were especially important for food and other uses to First Peoples and later for European explorers and settlers. Prior to the arrival of Europeans, wildlife use was regulated by land-tenure and resource-use rights of the various First Nations. Later, around 1822, the Hudson's Bay Company consolidated its control of wildlife and maintained a virtual monopoly until British Columbia became a colony in 1858. The first law governing ungulates in BC was the Act Providing for the Preservation of Game, introduced in 1859. It recognized the value of wildlife and established regulations governing its use, setting hunting seasons for several species, including deer. The act received various amendments, including some specifically for ungulates, until the Game Amendment Act of 1905 was passed to incorporate all the previous amendments and to introduce new ideas about wildlife management. The Game Act was itself replaced in 1966 with the Wildlife Act, developed with input from scientifically-trained wildlife managers. This act had a broader scope than just game species. It was later expanded in the Wildlife Act of 1982, which had two main goals: maintenance of biological diversity in the major biophysical regions of the province, and maintenance of wildlife numbers to meet the ecological, economic, social and recreation needs of society.

The European approach to management and conservation of wild animals originally focused almost exclusively on game species – those hunted for meat and sport. Ungulates are considered game animals and for 50 years following the province's first wildlife legislation, management was aimed primarily at the use of game animals with little or no regard to maintaining or enhancing their numbers, other than extensive predator control programs. This emphasis on use saw steadily declining populations of many wild species in parts of the province.

In the early 1900s, a new advance in wildlife management was promoted by President Theodore Roosevelt in the United States, and was also adopted in Canada. This new approach recognized

Figure 32. Roosevelt Elk in a regenerating clear-cut near Woss on Vancouver Island. Areas in the early successional stages of forest regrowth can provide good feeding conditions for ungulates. (D. Koshowski photograph.)

that natural resources were interrelated and that their wise use was a public responsibility. It also identified science as the management tool. In 1933, Aldo Leopold published his book *Game Management*, heralding the true beginning of scientifically-based wildlife management and conservation. The BC Game Commissioners of the day, James Cunningham and Frank Butler, were quick to adopt this new way of managing wildlife and fish resources. In 1950, they established an honorary board of scientific advisors consisting of Dr W.A. Clemens, Dr Wilson Duff and Dr Ian MacTaggart Cowan, all of the University of British Columbia. The commissioners also negotiated with the UBC's Department of Zoology to establish provincial laboratories for both fisheries and wildlife biology at the university.

The first scientific research on BC's ungulates was carried out by Dr Ian MacTaggart Cowan, who had been hired by the British Columbia Provincial Museum in 1935 as a biologist. Dr Cowan joined the Department of Zoology at the University of British Columbia in 1940 and established a research program devoted to wildlife biology. He supervised a large number of graduate students, many of whom later worked for the Provincial Wildlife Branch as managers and research scientists. One of his first students was James Hatter, who studied Moose in central British Columbia for his doctoral thesis. In 1947, Dr Hatter became the first game biologist with scientific training hired

Figure 33. A high-elevation forest in the Columbia Mountains north of Revelstoke. Subalpine forests with abundant arboreal lichens are vital for the survival of the Mountain ecotype of Woodland Caribou. (B. Allison photograph.)

by the province, and in 1962, went on to become the first Director of the Wildlife Branch. Many other biologists, mainly employees of the Wildlife Branch have contributed to management and conservation of BC's wild ungulates, including people such as John Bandy, Don Blood, Vernon (Bert) Brink, Kim Brunt, Ken Child, Dennis Demarchi, Ray Demarchi, Don Eastman, John Elliott, Fred Harper, Fay Hartman, Dave Hatler, Doug Heard, Daryll Hebert, Doug Janz, Pat Martin, Bruce McLellan, Harold Mitchell, Don Robinson, Ralph Ritcey, Ian Smith, Dave Spalding and Lawson Sugden, to name a few.

Today, the major challenges facing managers working with ungulates and other wildlife are directly related to large-scale human activities that affect the quality, fragmentation and availability of wildlife habitat. The four most important of these activities are forestry, agriculture, hydro-electric developments and mining. Forestry has major impacts on ungulate habitat, both beneficial and negative. For example, after logging, the early successional stages can increase the biomass of forages preferred by ungulates such as Black-tailed Deer and Elk (figure 32), and lead to population increases. Conversely, loss of high elevation old-growth forests with their abundant arboreal lichens (figure 33), and fragmentation of forest cover at lower elevations, are detrimental to Woodland Caribou. Loss or alienation of ungulate ranges, especially of their winter ranges, is a common result of agriculture and is a particular problem in the Okanagan and northeastern BC. In other areas of BC, ranching can create competition between wild and domestic ungulates for forage, and it can introduce diseases. The large reservoirs built for major hydro-electric projects eliminate habitat in valley bottoms – biologically, the most productive areas – which are also major ungulate winter ranges. The fourth activity, mining, also causes habitat loss mainly through open-pit mines either in low elevation winter ranges or at high elevations where reclamation can be extremely difficult. In addition, road development is usually associated with these activities and can have negative impacts on wild ungulates. Roads in back-country areas increase access to ungulates and other wildlife and can lead to local increases in both legal harvest and illegal killing. This presents a difficult conservation problem. Extensive road building for resource development and exploration led to population declines in some ungulate species up until the 1970s, when the problem was recognized and appropriate management implemented.

The overall scope of wildlife management and conservation in British Columbia has changed from its original narrow focus on game species, sport hunting and regulations, to a broader mandate involving the conservation of non-game vertebrates, threatened species management, habitat conservation, public education, non-consumptive use and research. Currently, the province is divided into eight administrative regions (figure 34), each subdivided into numerous units to aid in the management of ungulates and other wildlife resources. Each region is staffed by wildlife and habitat biologists responsible for wildlife conservation, hunting regulations, censuses and habitat management, among other tasks. Hunters in BC are required to buy a licence and the number of animals harvested has been regulated at various levels since the late 1890s. In 1974, an intensive management system – the Limited Entry Hunts (LEH) – was introduced as another means of controlling the age-and-sex-class harvest of a species in some areas.

The provincial government places all money received from hunting and fishing licences into general revenue, and the provincial cabinet decides on the amount of funds required for conservation and management. In 1980, several groups, including those representing hunters, anglers, guides and trappers, requested that a surcharge be levied on all licences provided that this revenue was devoted to conservation. This resulted in the establishment of the Habitat Conservation Fund. The fund's mandate includes the enhancement of fish and wildlife habitat, and the acquisition of land. The fund also supports some applied research and efforts to reintroduce species into their

Figure 34. Fish and Wildlife Management Regions. (Wildlife Branch, Ministry of Environment).

former range. The fund plays an important role in the conservation and management of BC's ungulates. It was renamed the Habitat Conservation Trust Fund in 1996, with a mandate broadened to include biodiversity.

The BC Ministry of Environment has a process to assign conservation priorities for species of animals and plants whose status in the province is not secure. The process names species on Yellow, Blue and Red lists. Species in the Yellow list are secure. Those named in the Blue or Red list need special attention if they are to remain part of the fauna or flora of British Columbia. The lists also provide a practical way to help make conservation and land-use decisions, and to prioritize research, inventory, management and protection activities. Since 1992, the lists have been coordinated with the international ranking system used by the BC Conservation Data Centre. Species are included in the Red List if they have been classified or are being considered for inclusion in the formal conservation designations of Extirpated, Endangered or Threatened. Not all Red-listed species are necessarily formally designated, but inclusion on this list indicates they are at risk and require investigation. The Blue List is for species at a lower level of risk and includes those considered to be vulnerable in the province, meaning that they are sensitive to human activities or natural events. In 2012, three ungulate subspecies were on the Red List – Woodland Caribou (Mountain and Boreal ecotypes), Plains Bison and Wood Bison – and four on the Blue List – Roosevelt Elk, Woodland Caribou (Northern ecotype), Rocky Mountain Bighorn Sheep and Dall's Sheep.

Protected areas within the province are also important for the conservation of ungulates. They include Canadian national parks, provincial parks (including wilderness parks), ecological reserves and nature conservancies. Hunting is prohibited in the majority of provincial parks and in all national parks and ecological reserves. Currently, 10.7 per cent of British Columbia is included in some form of provincial or federal protected area. Another land designation is the Wildlife Management Area (WMA) – these areas need intensive management because the land has significant wildlife habitat and population values. There are three WMAs important for ungulates: the grassland of the Dewdrop-Rosseau WMA (4,240 ha) near Kamloops, managed for Bighorn Sheep and Mule Deer; the East Side Columbia Lake WMA (6,886 ha) that has upland habitat used as winter range by several ungulate species; and the Hamlin Lakes WMA (30,572 ha) in the Kootenays that benefits Woodland Caribou and other species.

Selected References: Cowan 1987, Demarais and Krausman 2000, Demarchi and Demarchi 1987, Harper et al. 1994, Leopold 1942, Murray 1987, Robinson 1987.

Studying Hoofed Mammals

British Columbia's ungulates have been studied for many years, but there is still much we do not know, even about their basic biology. Most research is directed at their ecology (how they interact with their environment) to help managers conserve and manage the species. As a result, most studies have concentrated on wildlife food habits, habitat use, diseases and parasites, predators and other mortality factors, as well as on population dynamics and productivity. Some research questions that need solving are common to most if not all species, whereas others are specific to a certain species (e.g., the causes of die-offs in Bighorn Sheep).

Not surprisingly, a common research focus in BC has been the effects of forest practices on ungulates and other wildlife. Much effort has gone into this area and it remains a priority. One reason for continued research is that habitats and other environmental factors vary across the province, so what may be true for one area, may not apply equally elsewhere. And conditions in the same area can change, whether in the form of broad climate changes or through human activities. One example of concern to wildlife managers is the Woodland Caribou and its dependency on old-growth forests – what is happening to Caribou in the mountains of southeastern BC does not seem to apply directly to those in the northeast or central regions. At the same time, because of limited resources – researchers, funding and time – some important questions have received little study. Unfortunately, there are simply not enough resources to answer the major questions for all species at the same time.

The methods used to attain research and management goals not only reflect the questions identified for answering, but also depend on the logistical problems of studying a given species. Nothing beats direct observation, but it takes a lot of time and many of BC's hoofed mammals are not easy to observe. Some, like Roosevelt Elk and Black-tailed Deer, live in dense habitats that make direct observations difficult, and many of BC's ungulates travel long distances when they migrate between seasonal ranges during the year. Without prior information on the movements or ranges, it is extremely difficult to keep up with the animals and collect sufficient data. The usual research approach to overcome these problems is to use radio-telemetry.

Telemetry generally involves placing a special collar on an animal and then monitoring the signals transmitted by the radio built into the collar (figure 35). This of course means that the animal must first be captured safely and with minimal stress to it and the researchers. There are many different ways to catch ungulates and they range from tranquillizers to traps. If the animal is free-ranging, the operator either stalks it on foot or approaches in a helicopter, then shoots it with a tranquillizer dart. Alternatively, an animal can be trapped in fixed or portable traps baited with salt, hay, apple mash or another attractive food, or it can be captured in a small net fired from a special gun by an operator in a helicopter. This netting method is valuable if specific individuals need to be captured or if the animal's behaviour makes other trapping methods ineffective.

Researchers fasten a radio collar around the captured animal's neck, along with other identification devices (e.g., ear tags). They often weigh and measure the animal before releasing it. Monitoring the

Figure 35. An adult female Roosevelt Elk with her young in Strathcona Provincial Park on Vancouver Island. Researchers have fitted a VHS radio collar around her neck so they can monitor her movements.

signals transmitted by the radio collars and relocating the animals is done either from the air using helicopters or fixed-wing aircraft, or from the ground using a hand-held directional antenna (figure 36) to determine the direction of the signal and then taking a compass bearing. On the ground, bearings are taken from at least three different stations and plotted on a map to estimate the animal's location. The location coordinates are entered into a computer database for analyses of animal movements, home range sizes and habitat use. Recent technological advances allow the collars to do most of the work with the help of satellites. Either the satellite monitors the signals from the collar, or the collar monitors signals broadcast by Global Positioning System (GPS) satellites at regular intervals and stores the animal's locations. The data are either transmitted to a central station by satellite and then to the researcher, or the GPS collar releases its stored location information and other data when the researcher sends a radio signal to the collar. Among the many advantages of GPS collars, besides the immensely large sample size, is the recording of data throughout the full 24-hour day and during periods of inclement weather when conventional ground or aerial relocation methods are difficult if not impossible.

Figure 36. Researchers use a portable antenna and a radio receiver, either from the ground or the air, to follow the movements of hoofed mammals fitted with radio collars (see figure 35). Recent developments with GPS collars provide more effective data.

The BC Ministry of Environment has standards for animal inventories conducted with provincial funds. Standard forms for recording data during the capture and handling of ungulates require a researcher to collect a series of morphometric measurements, including body weight, chest girth, total length, tail length, shoulder height and ear length. These measurements will be housed in a central provincial database where they will be made available to interested scientists.

The task of studying BC's ungulates is not limited to professional biologists. Amateurs can and have made important contributions to our understanding of these animals, especially about their behaviour and ecology. If you wish to study ungulates, start by reading as much as you can of what is already known – this will help you find what is still unknown so you can direct your research efforts accordingly. No matter what topic you study, keep meticulous field notes. Include the date, time, place and weather, along with details of what you saw or found. Examples of useful quantitative observations you can record are group sizes and their age-sex composition, and ratios such as the number of adult males to adult females or the number of young to adult females. A good pair of binoculars and even a spotting scope are useful for studying ungulates, especially in areas where they are hunted. A small portable tape recorder is also helpful to record notes while in the field; just remember to check it periodically to ensure that it is still recording.

But even if you don't want to carry out a scientific study, you can enjoy our native ungulates in other ways by simply watching or photographing them. British Columbians are fortunate because this province is especially rich in wild ungulates, and there are lots of opportunities for seeing them in their spectacular surroundings. The greatest diversity of ungulate species is in northeastern BC, but access can be limited. Fortunately, there are many other regions with better viewing opportunities, such as the East Kootenay in the southeast (figure 37). Look in the Range section of each species account for suggestions on good places to look for ungulates. There is also a useful website hosted by the BC Ministry of Forests (http://www.env.gov.bc.ca/fw/wildlife/viewing/), which also provides links for detailed information by region. Besides patience, it also pays to find out as much as you can about an ungulate's habits before you go out to find or photograph it. That way, you will increase your chances of success and also of understanding what you see.

Hoofed mammals' unique vision means that they can be difficult to study or photograph, even in national parks where they may be quite used to people. You will need patience to get within camera range – don't try to sneak up on them, especially species living in the open. It's better to walk obliquely toward them, on a zigzag path, without looking directly at them. It also helps to stop frequently and look away, especially if the animal seems to have spotted you. For forest dwelling species, it is probably best to sit and wait quietly near an area known or likely to be used by the species you are interested in. But be careful not to get too close to many of our native ungulates, not only because many are large animals, but because they can be quite aggressive. Females will often defend their young, and males can be particularly aggressive during the rut. Even though their weapons are primarily for fighting mating rivals, they will not hesitate to use them against a human who gets too close. In protected areas, such as a national park, animals may be tolerant of humans and may even appear tame, but they can be particularly dangerous because they have little fear of humans. And remember, should you find a deer fawn on its own, it is almost certainly not abandoned. It is normal for their mothers to leave them during the day while they feed elsewhere.

It can be extremely rewarding to spend a few minutes watching wild ungulates, even if they are just feeding or resting. You can see how they select their food, choosing one patch of plants over another,

Figure 37. The East Kootenay region of southeastern British Columbia can provide some of the best opportunities in the province to see many of our native species of ungulates.

how they grasp the food with their lips or tongue, and if resting, how they ruminate, watching the lump of food travelling up and down the throat between mouth and stomach. You might also see social interactions between group members, or see a young come to its mother and suckle. If you are lucky enough to observe hoofed mammals during the mating season, you will usually see lots more action. Males are particularly active, and depending on their mating system, may be defending a group of females from other males, or vigorously courting a female using a variety of behaviour patterns.

The skulls of many species, especially of adult males, are quite robust, so you can often find them in areas where ungulates spend the winter and early spring. The key to the skulls in this book will help you to identify any that you find. You can also use the information presented earlier in this section to determine what its feeding habits were. If you want to estimate its age, you will need other reference materials to help interpret tooth eruption sequence and wear for all species, or to employ other methods. The Wildlife Society's two-volume *The Wildlife*

Techniques Manual (Silvy 2012) is particularly helpful for estimating the age-sex class of a specimen. If you pick up any wildlife parts, such as skulls (but not cast-off antlers), and do not wish to leave them for someone else to enjoy, you must report to the BC Wildlife Branch to obtain a possession permit; for any wild sheep skull, no matter how it is obtained, the horns must be pinned with an identification plug by a wildlife official.

Selected References: Demarais and Krausman 2000, Silvy 2012, Stelfox 1993, Wareham 1991.

Tracks and Signs

Ungulates leave distinctive tracks – the bi-lobed shape of cloven hooves is unique to Artiodactyls, and the horse's single round hoof-mark is also easy to recognize. Unfortunately, it is not always easy to identify the species of ungulates by their tracks because of close similarities. For example, the tracks of the three subspecies of Mule Deer and those of White-tailed Deer are very difficult to tell apart; it is also hard to separate Domestic Cattle tracks from those of Bison. But some species are easier to identify. The Caribou's tracks are characteristically round, being formed by two sausage-shaped main hooves that often leave a track wider than it is long. The Caribou's dew claws almost always leave an impression; for other species, the substrate usually has to be quite soft for the animal to sink far enough in to leave an imprint of its lateral hooves.

The illustrations of ungulate tracks (figure 38) may help you identify some of the tracks you find. But you may also need the location, distribution and habitat type to help you decide what species was likely to have left the tracks. The main pair of toes are flexible, so hoof prints can vary somewhat from those shown in the illustrations, changing with the conditions of the ground, if the animal is running or walking, or if it is moving up or down hill. For example, when moving downhill, the tips of the hooves are often splayed apart. And many species place the hind foot over the track of the forefoot when walking, obliterating a clear print of a single hoof.

Besides tracks, ungulates leave other signs of their presence. Many deer species rub and scrape their antlers against vegetation, leaving tell-tale scrapes or blunt grooves running vertically down the trunks of small trees, or clumps of broken branches in shrubs. Signs of browsing on shrubs and trees are also easy to see (figure 39); they are recognizable not only by the clipped-off tips of shoots, but also by the plant's

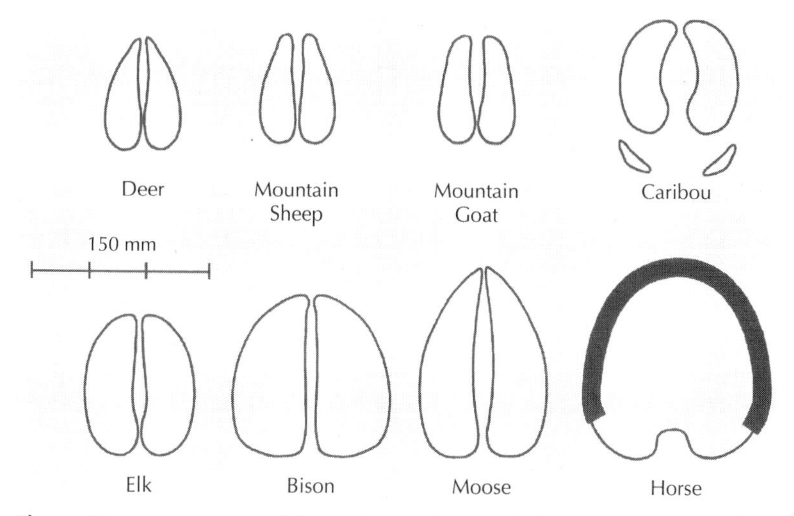

Figure 38. A comparison of the tracks of adult hoofed mammals in BC, all drawn to approximately the same scale. Tracks will vary depending on the condition of the ground, the speed of the animal, the topography and other factors. The dark rim on the horse track indicates the shoe impression, which will be absent in the tracks of feral horses.

Figure 39. Signs of feeding by ungulates: Fireweed (left) topped off by a Roosevelt Elk and Red-osier Dogwood (right) browsed by one or more Moose. In both cases, removal of the main stem stimulates the growth of lateral shoots, but heavy browsing over several years can kill a shrub.

production of many lateral shoots. Initially, this results in an increase in available food for the ungulate, because instead of a single main shoot, there are many tender lateral shoots; but severe, long-term browsing leads to overuse, called hedging, that can eventually kill the plant. Species such as Elk, Moose and Bison wallow and may dig shallow depressions in the ground. Bison may use the same wallows over many years, so these wallows can become very large in diameter. The Species Accounts give more details on characteristic signs left by different species in BC.

Like tracks, the feces or droppings of ungulates as a group are easy to recognize, but separating the different species is even more problematic. One reason is that pellets vary in size not only between species but also with an animal's age. Shape is also affected by the type of food eaten. In spring, plants have a relatively high moisture content and the soft feces can resemble either a stack of small pancakes or miniature cow-pats, while at other times of the year, when food is mature and usually drier, the droppings can be lozenge-shaped. Freshness of the fecal pellet also affects shape because drying can obliterate or otherwise alter small identifying features. For these reasons, there is no particularly reliable identification guide for ungulate feces, although some books are useful, such as those referenced here.

Selected References: Murie 1975, Rezendes 1992, Stelfox 1993.

Taxonomy and Nomenclature

To describe, study and understand living organisms, we need to give them names. The most useful systems for naming organisms are those that indicate relationships between the different types, and are ones that can be used by anyone regardless of their native language. Formal classification systems of living organisms have been around at least since Aristotle's time. The one we use today was developed by the Swedish botanist Carl Linnaeus in 1758. His method of nomenclature was based on a binomial system in which an organism is described by two names, usually italicized – the genus and the specific name. This formal description sometimes includes the name of the person who gave the first scientific name, along with the year it was described. For example, the scientific (binomial) name for the species Thinhorn Sheep is *Ovis dalli*, Nelson 1884. This means that this wild sheep belongs to the genus *Ovis*, has the specific name *dalli* and was first described and named by E.W. Nelson in 1884. Today, we also use three names or a trinomial to describe races or geographic subdivisions of a species. The Thinhorn Sheep has two subspecies in British Columbia: *Ovis Dalli dalli* (Dall's Sheep) and *Ovis dalli stonei* (Stone's Sheep). By using Latin names, the Linnaen classification system can be used around the world.

Originally, biological classification systems, including Linnaeus's, were based on physical similarities, but since Darwin published his *Origin of Species*, the biological classification system has attempted to reflect evolutionary (phylogenetic) relationships among organisms. With the recent developments in DNA techniques, we can base classification on firm genetic relationships. But as recent research on North American deer shows, this task is proving challenging and will not answer all our questions.

In addition to the genus and specific levels, the Linnaen system also provides a hierarchy of other taxonomic categories. For example, at one of the broadest levels, all warm-blooded animals with hair are included in the Class Mammalia. The next level down after class is order, and as we saw earlier, hoofed mammals are divided into two orders: (Cet)Artiodactyla and Perissodactyla. Below order are suborders, families, sometimes subfamilies and tribes, and finally the levels of genus, species and subspecies. The checklist of ungulates, following this section, illustrates some of these levels.

The scientific names used in this book follow *Mammal Species of the World* (Wilson and Reeder 1993). Taxonomy is a dynamic discipline – new information or new ideas appear and names can be changed. Hence, not all biologists use the same names as in this reference, although the differences in nomenclature are usually minor. The same cannot be said of common names. Unlike the birds, there is no standard list of common names for ungulates or other mammals. Common names for mammals vary from place to place, and many species and subspecies have more than one. For example, Shiras' Moose is also called Yellowstone Moose and Wyoming Moose, and Roosevelt Elk is also known as Vancouver Island Elk and Olympic Elk. Most of the common names used in this book follow the BC Ministry of Environment's online publication *The Vertebrates of British Columbia: Scientific and English Names* (http://www.ilmb.gov.bc.ca/risc/pubs/tebiodiv/vertebratev3/taxa_ml_v3_02all.pdf). Exceptions are where the name used here is the one given with the original description or if there has been a more recent taxonomic revision. Frequently used alternative common names are included in the Species Accounts.

The concept of ecotype, previously applied mainly to plants, has been used recently for some mammal groups in BC to recognize distinct ecological and behavioural differences within a species or subspecies. In part, use of this concept may be an attempt to avoid problems that sometimes arise when recognizing or defining subspecies. For BC ungulates, the ecotype concept has been applied to Woodland Caribou and Mountain Goat populations to reflect differences in their environment and in the animals' responses to these differences. More recently, there has been discussion about whether Moose in BC should be subdivided into a number of ecotypes. It might be argued that Rocky Mountain and California Bighorn Sheep, formerly recognized as subspecies, may similarly be considered ecotypes in BC.

Selected References: Cronin 1992, Nagorsen 1990, Wilson and Reeder 1993.

C-1. Dominant Bighorn Sheep use the same patterns as they would for courting a female when they interact with a subordinate male. Here, the dominant male combines a nose twist with a kick.

C-2. Deer avoid using their antlers to fight while they are covered in velvet and still growing.

C-3. A male Rocky Mountain Elk in summer with a large set of growing antlers.

C-4. A deer with hardened antlers has begun scraping off the velvet (see page 37).

C-5. Large ears, a sensitive nose and eyes on the sides of the head help ungulates detect predators.

C-6. Young Bison have an orange coat until they are about three months old, after which it gradually changes to the typical brown.

C-7. A mother Mule Deer leads her fawns to a general hiding area, where they will select their own hiding places (see page 49).

C-8. Like all female ungulates, this mother deer smells and licks the anus of her fawn as it comes to suckle, probably to check its identity.

C-9. The long legs of a Moose allow it to move relatively easily through deep snow to reach forage and escape from predators.

C-10. Snow poses problems for grazers, making low-growing forage more difficult to find and eat.

C-11. Mule Dear finding a little shelter from the wind in a grove of bare aspen saplings in winter.

C-12. A female Mountain Goat starting to shed her thick winter coat.

C-13. A Black-billed Magpie has landed on a female Rocky Mountain Bighorn Sheep to pick off the ticks. The sheep no doubt welcomes this short-term symbiotic relationship.

C-14. This White-tailed Deer signals alarm by raising its tail to expose the white underfur, a behaviour called flagging (see page 151).

C-15. A male Bison roars like a lion when challenging other males during a rut.

C-16. The mass of hair on this male's forehead may help cushion the blow during a rutting fight.

C-17. An adult female Black-tailed Deer strikes a male yearling with her foot (see page 138). Her drawn-back ears is also a sign of aggression.

C-18. When competing for a female, adult male Bison butt heads and wrestle head to head, attempting to gore their opponent.

C-19. Although males vie aggressively for mates, in some species that form tending pairs, they often do not compete when testing females for signs of heat (see pages 47–48 and 209).

C-20. European Fallow Deer engaging in lekking behaviour, where males congregate and hold very small territories on which they attempt to mate with females (see pages 113–14).

CHECKLIST

Orders, suborders, families, subfamilies, tribes and genera are arranged according to generally accepted phylogenetic order. Species and subspecies within a genus are listed alphabetically by scientific name. This list includes recently extinct species and feral species of domestic livestock.

Order Artiodactyla: Even-toed ungulates

Family Cervidae: Deer

Subfamily Cervinae: Old World Deer

Cervus elaphus Linnaeus	Elk
Cervus elaphus nelsoni Bailey	Rocky Mountain Elk
Cervus elaphus roosevelti Merriam	Roosevelt Elk
Dama dama dama Linnaeus	European Fallow Deer*

Subfamily Odocoileinae: New World Deer

Alces alces (Linnaeus)	Moose
Alces alces andersoni Peterson	Northwestern Moose
Alces alces gigas Miller	Alaskan Moose
Alces alces shirasi Nelson	Shiras' Moose

* This non-native wild species introduced to British Columbia is included in the Species Accounts.

Odocoileus hemionus (Rafinesque) Mule Deer
Odocoileus hemionus columbianus (Richardson)
Columbian Black-tailed Deer
Odocoileus hemionus hemionus (Rafinesque)
Rocky Mountain Mule Deer
Odocoileus hemionus sitkensis (Merriam)
Sitka Black-tailed Deer

Odocoileus virginianus (Zimmermann)
White-tailed Deer
Odocoileus virginianus dacotensis Goldman and Kellog
Dakota White-tailed Deer
Odocoileus virginianus ochrourus (Bailey)
Northwestern White-tailed Deer

Rangifer tarandus Linnaeus Caribou
Rangifer tarandus caribou Gmelin Woodland Caribou
Rangifer tarandus dawsoni Seton-Thompson
Dawson's Caribou (extinct)

Family Bovidae: Hollow-horned ruminants

Subfamily Bovinae: Bison and cattle

Tribe Bovini:
Bison bison (Linnaeus) Bison
Bison bison athabascae Rhoads Wood Bison
Bison bison bison (Linnaeus) Plains Bison

Bos taurus Linnaeus Domestic Cattle*

Subfamily Caprinae: Sheep, goats, goat-antelopes, muskoxen and takin

Tribe Rupicaprini: Goat-antelopes
Oreamnos americanus Blainville Mountain Goat

Tribe Caprini: Sheep and goats

Capra hircus Linnaeus — Domestic Goat*

Ovis aries Linnaeus — Domestic Sheep*

Ovis canadensis (Shaw) — Bighorn Sheep
Ovis canadensis canadensis (Shaw) — Rocky Mountain Bighorn Sheep

Ovis dalli Nelson — Thinhorn Sheep
Ovis dalli dalli Nelson — Dall's Sheep
Ovis dalli stonei J. A. Allen — Stone's Sheep

Order Perissodactyla: Odd-toed ungulates

Family Equidae: Horses, asses, onagers and zebras

Equus caballus Linnaeus — Domestic Horse*

* Feral populations of domestic livestock in British Columbia are not included in the Species Accounts.

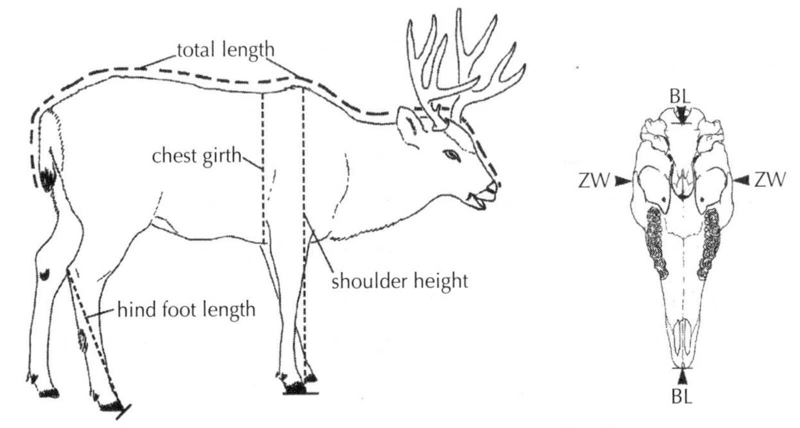

Figure 40. Standard body and skull measurements used in the species accounts. The two skull measurements are also used in the Key to the Skulls.

Weight. Weigh the whole animal in kilograms. If the animal is dead, weigh it before it is bled.

Total Length. Measure when the animal is laid on its side with the head, neck and tail stretched out. If this is not possible, take the measurement along the length of the animal as shown above.

Tail Vertebrae (not shown above). With the tail raised at a right angle to the back (vertically), measure from the base of the tail (its junction with the back) to the tip of the last vertebra.

Hind Foot Length. With the lower part of the hind leg extended, measure from the proximal end of the heel (calcaneus) to the tip of the hoof.

Shoulder Height. With the animal lying on its side and the front leg straight, measure the straight-line distance from the tip of the hoof to the mid-line of the back just above the shoulder blade.

Ear Length (not shown above). Measure the straight-line distance from the notch at the bottom of the ear opening to the tip of the ear cartilage.

Chest Girth. Measure the circumference of the chest immediately behind the elbow of the front leg.

Skull Length (Basilar Length – BL). Measure from the tip of the premaxillae to the anterior edge of the foramen magnum.

Skull Width (Zygomatic Width – ZW). Measure the greatest width of the skull across the outside of the zygomatic arch.

IDENTIFICATION KEYS

The two keys that follow will help identify the hoofed mammals of British Columbia. One is for live animals seen in the field and the other is for cleaned skulls. Both keys are dichotomous, and the procedure for identifying whole animals or skulls is the same. The diagnostic features are arranged in paired statements that offer two mutually exclusive choices (a or b). I have avoided using subjective traits (e.g., slightly darker than) and instead use the presence or absence of features or measurements. To identify an ungulate, start with the first pair of statements and select the option that best describes the animal or skull. The number at the end of the option you selected directs you to another pair of options in the key. Repeat these steps and you should eventually arrive at an identification.

The keys allow identification of the species and some subspecies. Once you have arrived at an identification, consult the appropriate species account to see if it is consistent with your determination from the key. The descriptions in the accounts will also help identify the subspecies. In some cases, I have provided what may seem to be extraneous identifying characters, but because skulls found in the field are often broken, it is sometimes useful to have extra distinguishing criteria.

In addition to total body weight, five standard museum measurements (figure 40) are usually collected for ungulates; and for wildlife studies, a sixth – chest girth – is also frequently taken because it can be a useful estimator of body weight. For the purposes of this key, two skull measurements are also defined. Although not used in the identification keys, weights and measurements of adult specimens are provided in the Species Accounts. In this handbook, measurements are reported in metric units. The symbol > indicates greater than and < indicates less than. For example, "tail >150 mm" means that the tail is more than 150 mm long.

Key to Live Animals

I designed this key to help identify living animals in the field using features that can be readily distinguished by eye with the aid of binoculars or a spotting scope where necessary. The key applies to adult animals of both sexes. The drawings of live animals accompanying the species accounts will also help identify the species. For some, locality may be the only means of distinguishing the subspecies: Moose in southeastern British Columbia are almost certainly Shiras' Moose; Elk on Vancouver Island are Roosevelt Elk; and Black-tailed Deer on Vancouver Island and south of Rivers Inlet along the coast are Columbian Black-tailed Deer. For subspecies identification consult the Description section in the appropriate species account. Although the most likely places to see free-ranging European Fallow Deer are on James and Sidney islands, the key allows identification of the three main colour phases: typical (spotted), black or dark brown, and white.

1a. Large body, at least as big as a saddle horse 2

1b. Small to medium body, no larger than a small pony 4

2a. Dark to reddish-brown body with short front legs; large square head with shaggy hair between horns and a distinct beard on the chin; shoulders higher than the head; short black horns that curve out and then back toward the head; long, shaggy hair on the shoulders and down the front legs .... Bison (*Bison bison*)

2b. Dark or reddish-brown body; long, slender legs; no horns, but may have antlers; no shaggy hair on the head or legs; no beard 3

3a. Dark brown-black body, with lighter coloured long, thin legs and no obvious rump patch; elongated head with a bulbous nose; small flap of skin hanging from the neck; if present, antlers are palm-shaped and extend laterally........... Moose (*Alces alces*)

3b. Reddish-brown body with dark-brown legs and neck; large

cream-coloured rump patch; antlers, if present, consist of a round main beam extending up from the head, with up to seven tines on each side, brown in colour with light tips
.... Elk (*Cervus elaphus*)

4a. White or cream-coloured body... 5

4b. Brown or darker body... 7

5a. Cream or pale-fawn body; tail >150 mm long; no horns; antlers, if present, rise up and backward from the head, and are palm-shaped toward the end; no metatarsal glands
.... European Fallow Deer (*Dama dama dama*)

5b. White or light cream-coloured body; tail <150 mm long; horns always present... 6

6a. Black horns, cylindrical in cross-section, smooth and sharply pointed, curving slightly back from the head; slender, pointed ears, at least half the length of the horns; winter coat has shaggy hair, a short beard and long hair on the legs to about 100 mm above the hooves; tail is white, > 100 mm long and covered in medium-length hairs .... Mountain Goat (*Oreamnos americanus*)

6b. Horns are brown and either elliptical in cross-section, short and curved slightly back with blunt tips, or triangular in cross-section, massive and curled into a tapered spiral; no beard; tail is usually < 100 mm and covered in short hairs
.... Dall's Sheep (*Ovis dalli dalli*)

7a. Either antlers or no other appendages on the head; large, wide ears 8

7b. Horns always present; small ears; large white rump patch; short black tail.. 13

8a. Reddish-brown body with horizontal rows of distinct white spots in summer that may be less obvious in winter; light-coloured belly and insides of the legs; tail covered in long dark-brown or black hairs; small white rump patch beneath the tail is bordered on both sides by a narrow but distinct band of black hairs; antlers, if present, rise up from head and are palm-shaped toward the ends; no metatarsal glands
.... European Fallow Deer (*Dama dama dama*)

8b. Body lacks horizontal rows of white spots.............................. 9

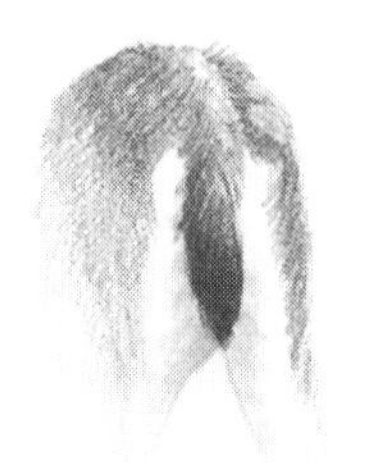
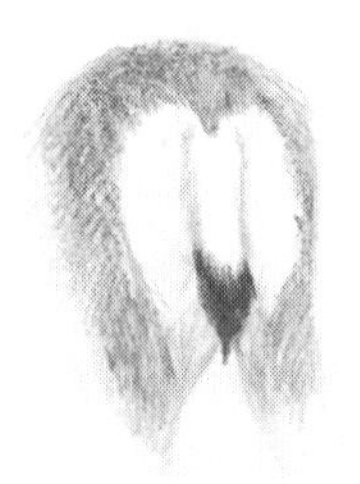

Figure 41. A comparison of the rump patches and tails to aid identification of Black-tailed Deer (left), Rocky Mountain Mule Deer (centre) and White-tailed Deer (right).

9a. Dark, almost chocolate-brown body in summer that can be lighter grey or brown in winter; neck and belly are lighter coloured and there may be lighter areas on the sides; the short white tail is surrounded by a light-coloured rump patch; a thin, distinct ring of white hair runs along the upper edge of each main hoof; square, blunt nose; antlers, if present, have a smooth surface and the tines may be branched and flat
.... Woodland Caribou (*Rangifer tarandus caribou*)

9b. Uniformly brown body, including the sides of the neck; no ring of white hairs above the hooves 10

10a. Uniformly dark, almost black-brown body, including the belly; no white hairs around the tail; no white throat patch; sometimes a faint indication of light spots horizontally along body; antlers, if present, rise up and backward from head and are palm-shaped toward the ends; no metatarsal glands
.... European Fallow Deer (*Dama dama dama*)

10b. Reddish-brown to greyish-brown body; white throat patch; white hairs around the tail or a white rump patch 11

11a. No dark cap or patch of hair between the ears; a white ring around the eyes is sometimes visible; no white rump patch; tail about 300 mm long and covered with long brown hairs, with white hairs at the margins and on the underside (figure 41); metatarsal gland is about 30 mm long, with white hairs; antlers, if present, have a main antler beam horizontally oriented with unforked tines branching from it (figure 42)
.... White-tailed Deer (*Odocoileus virginianus*)

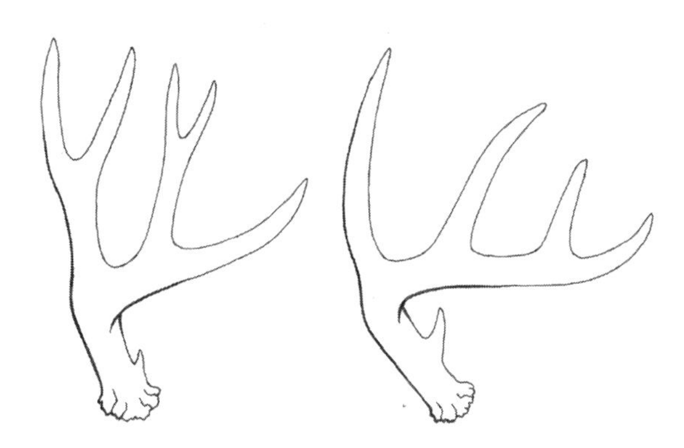

Figure 42. The antlers of Mule Deer (left) usually have forked tines extending from the main beam, unlike the single tines of White-tailed Deer (right).

11b. Dark cap or patch of hair on the forehead between the ears; metatarsal gland > 50 mm long, without white hairs in the centre (figure 41); if present, the main antler beam rises upward from head and the tines branching from it are usually forked (figure 42) 12

12a. Large white rump patch surrounding a slender, short-haired white tail with a black tip (figure 41)
.....Rocky Mountain Mule Deer (*Odocoileus hemionus hemionus*)

12b. Small white rump patch surrounding dark brown tail with a black tip (figure 41)
.................Black-tailed Deer (*Odocoileus hemionus columbianus* or *Odocoileus hemionus sitkensis*)

13a. Stocky, light- to dark-brown body and neck; large white rump patch and short black tail; horns are either short and curved slightly back with blunt tips, or massive, tapered and curled into a spiral.................................. Bighorn Sheep (*Ovis canadensis*)

13b. Stocky, grey to black coat, with the neck sometimes lighter; large white rump patch and short black tail; horns are either short and curved slightly back with blunt tips, or massive, tapered and curled into a spiral
....Stone's Sheep (*Ovis dalli stonei*)

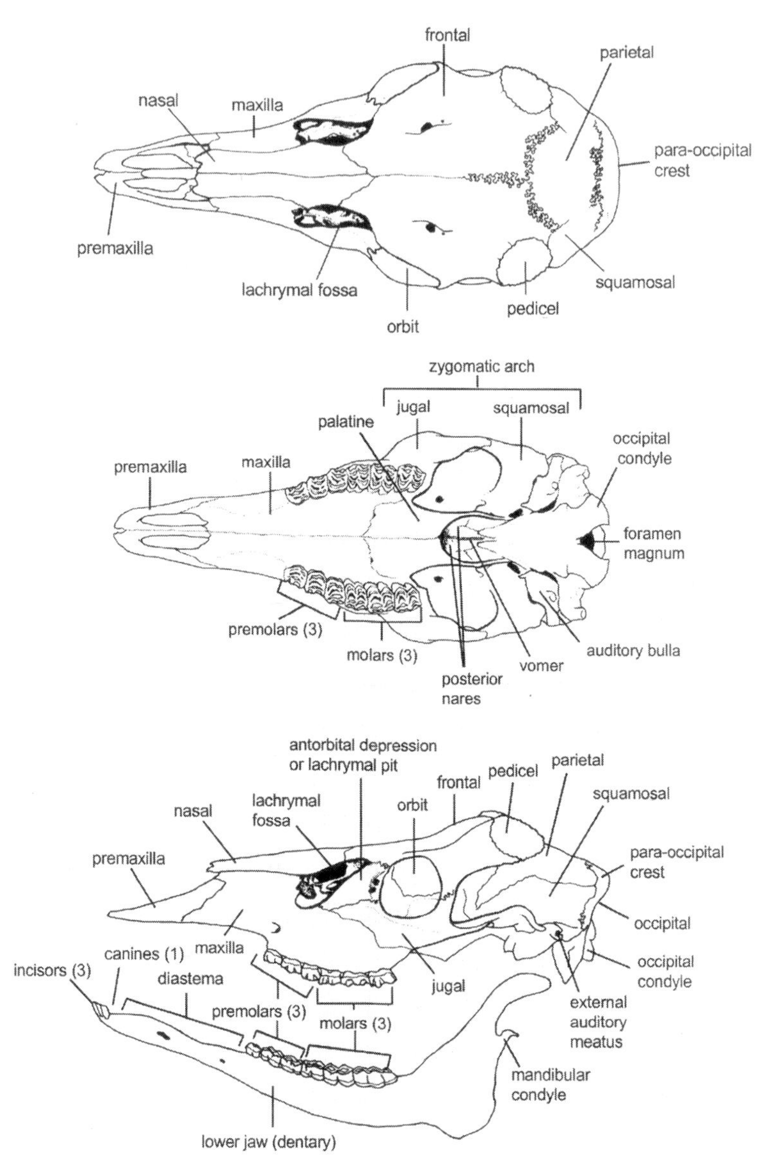

Figure 43. Dorsal (top), ventral (middle) and lateral (bottom) views of the skull of an adult male White-tailed Deer showing the main bones of the ungulate skull, most of which are referred to in the text.

Key to Skulls

This key relies on cranial and dental features. I developed it to use with cleaned, complete skulls of adult animals in museum collections (figure 43). It should also be useful for skulls picked up in the field, although damage and weathering will make identification more difficult. Bears and Grey Wolves feeding on ungulate carcasses often chew the front part of the head, so the premaxilla region may be missing. In some species, female and juvenile skulls are more fragile than those of adult males, making them more likely to be damaged and less likely to be found. This key includes features that should help distinguish species regardless of sex, but those of juveniles may be difficult to identify. Juvenile specimens, unlike adults, do not have a complete set of permanent teeth. If the lower jaw is present, the third premolar from the front can be used to distinguish juveniles from adults. This third premolar has three cusps in juveniles, while the adult's has only two (figure 44). Juveniles can also be recognized if any of their teeth are still erupting (figure 44). Only the skulls of adult female Bighorn Sheep and Thinhorn Sheep cannot be easily separated from each other. I have also provided identification criteria for the main domestic livestock, because skulls of these ungulates are sometimes found in the field. You will require a tape measure or calipers for measuring skull and horn bone-core lengths. The drawings of the skulls accompanying the species accounts will also help identify the species.

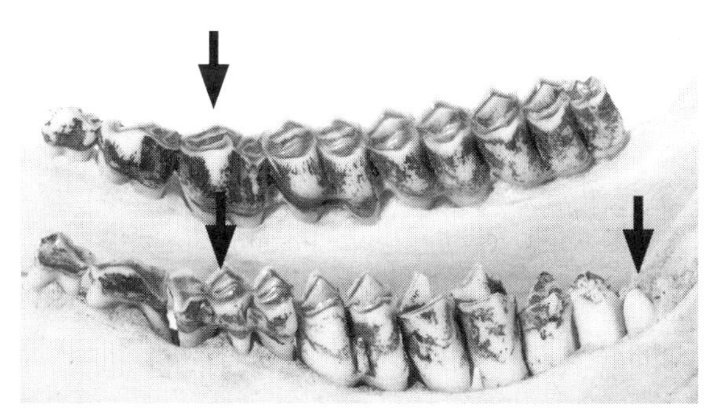

Figure 44. In juvenile ungulates of all species in BC, the third cheek tooth from the front in the lower jaw (bottom), the deciduous premolar, has three cusps (segments), whereas in the adult (top), the permanent premolar has two cusps. The last tooth (right) in the bottom jaw is still erupting, another indication of a juvenile animal.

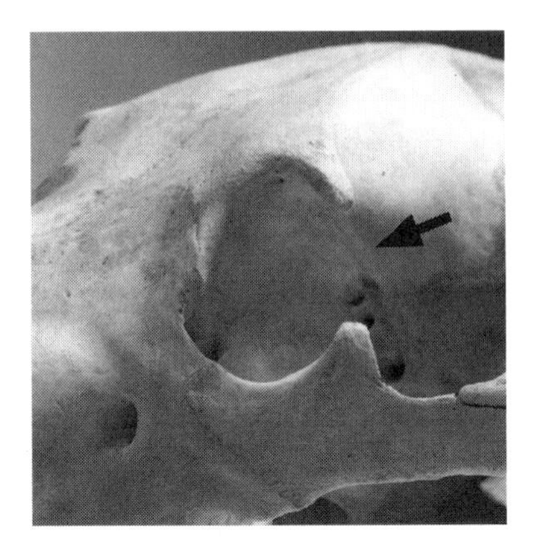

Figure 45. The skull of a carnivore, showing the incomplete orbital rim. The orbit in all ungulates' skulls (except that of pigs), forms a complete circle (see figure 46).

The first step in identifying a skull is to determine if it belonged to an ungulate. The skull of every adult ungulate species in British Columbia is more than 200 mm long, and what is unique to skulls of all ungulate species, except pigs (see 2b, below) is that the orbital rim forms a complete circle. In similar sized carnivore skulls, for example, the posterior rim of the orbit is incomplete (figure 45). Once this initial identification is made, move on to the key to find what species of ungulate the skull belongs to.

1a. Upper incisors present.. 2

1b. Upper incisors absent... 3

2a. Upper incisors massive and columnar; canines either small or absent (females); lower jaw massive; premolars and molars are hypsodont and lophodont; small sagittal crest ....Horse or Ass (*Equus* species)

2b. Upper incisors small and peg-like; large upper and lower canines directed outward and upward, triangular in cross-section with sharp edges; premolars and molars are brachydont and bunodont; well-developed sagittal crest and enlarged occipital region; incomplete orbital rim ....Domestic Pig (*Sus scrofa*)

3a. Large lachrymal vacuity between the nasal and lachrymal bones (figure 46, A); two lachrymal foramina on the anterior edge of the orbit (figure 46, B); pedicels or antlers may be present; horns or bone cores absent 4

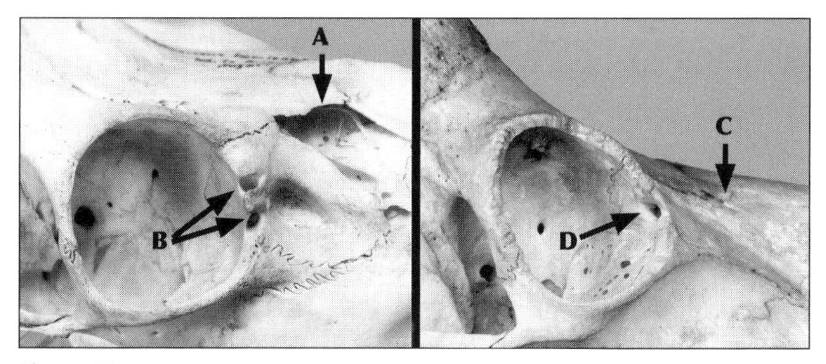

Figure 46.

3b. No lachrymal vacuity (figure 46, C); a single lachrymal foramen on the anterior rim of the orbit or just inside the front of the orbit (figure 46, D); horns or bone cores usually present; pedicels or antlers absent 9

4a. Upper canines present 5

4b. Upper canines absent 6

5a. Vomer divides the posterior nares (figure 47); premaxilla does not contact the nasal bone and has a prominent angle (figure 47); canines small, pointed and usually do not project below the premaxilla.......... Woodland Caribou (*Rangifer tarandus caribou*)

5b. Vomer does not divide the posterior nares (figure 47); premaxilla contacts the nasal bone and lacks a prominent angle (figure 47); canines squat, rounded and project well below the premaxilla.. Elk (*Cervus elaphus*)

Figure 47.

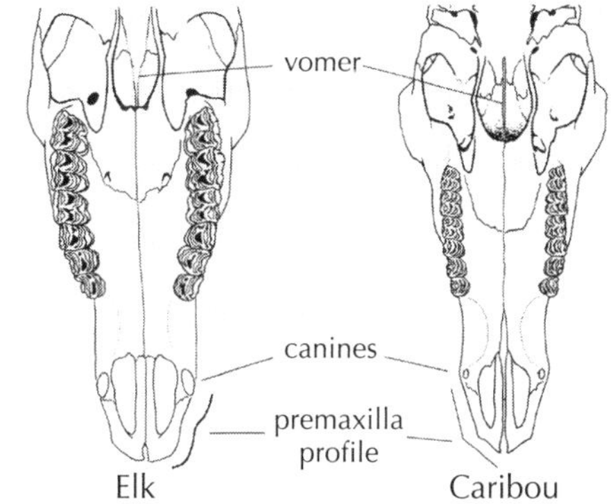

vomer
canines
premaxilla profile
Elk
Caribou

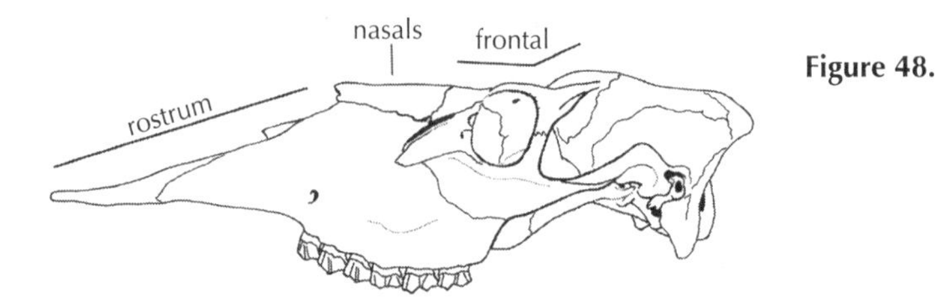

Figure 48.

6a. Basilar length > 500 mm; premaxillae and maxillae are long (> 200 mm) and slender; a small depression is followed by an obvious bump on the mid-line of the frontals between the antlers or pedicels in males, and just behind the orbits in females (figure 48) .... Moose (*Alces alces*)

6b. Basilar length < 500 mm; no depression and associated bump on the mid-line of the frontals 7

7a. Vomer divides the posterior nares into two parts (figure 49, top) 8

7b. Vomer does not divide the posterior nares (figure 49, bottom) .... European Fallow Deer (*Dama dama dama*)

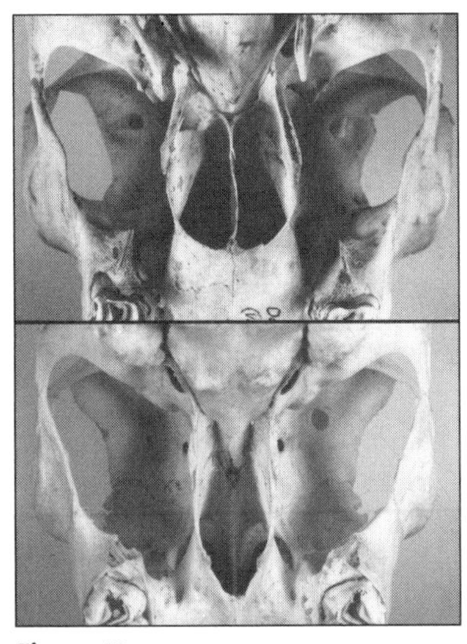

Figure 49.

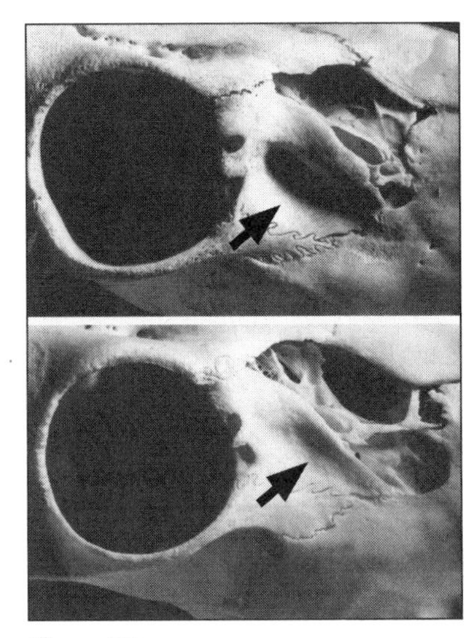

Figure 50.

8a. Shallow lachrymal depression or antorbital pit (figure 50, bottom); width of the auditory bulla is about equal to the length of the bony tube leading to the meatus; parieto-frontal suture curved toward the anterior edge .... White-tailed Deer (*Odocoïleus virginianus*)

8b. Deep antorbital (lachrymal) pit (figure 50, top); parieto-frontal suture more or less straight .... Mule Deer (*Odocoïleus hemionus*)

9a. Horns or bone cores present 10

9b. Horns or bone cores absent 17

10a. Basilar length > 350 mm; zygomatic width > 170 mm; bone core projects laterally from the sides of the skull 11

10b. Basilar length < 350 mm; zygomatic width < 170 mm; bone core projects up and backward from the top of the skull....... 12

11a. Wide frontals hide the zygomatic arches in dorsal view (figure 51, a); base of the bone core is anterior to the para-occipital crest (figure 51, b); para-occipital crest in dorsal view is slightly convex or has a single central bulge (figure 51, c); skulls of both sexes in dorsal view are clearly tapered from posterior to anterior... Bison (*Bison bison*)

11b. Zygomatic arch visible (figure 51, a) in dorsal view behind narrow frontals; base of the bone core extends directly from the para-occipital crest (figure 51, b); para-occipital crest in dorsal view bulges on either side of the centre line (figure 51, c); skull in dorsal view is slightly tapered from posterior to anterior .... Domestic Cattle (*Bos taurus*)

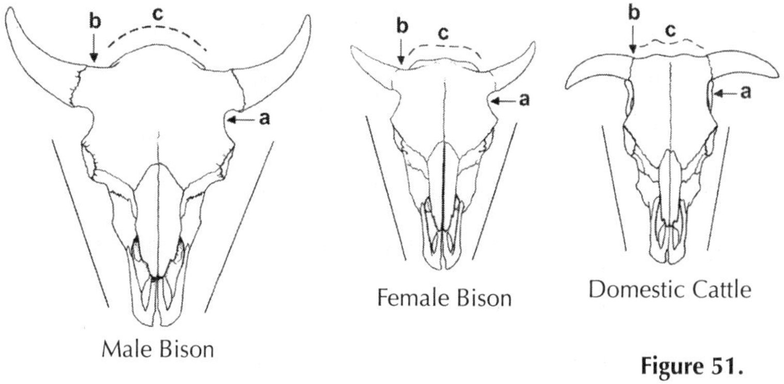

Male Bison

Female Bison

Domestic Cattle

Figure 51.

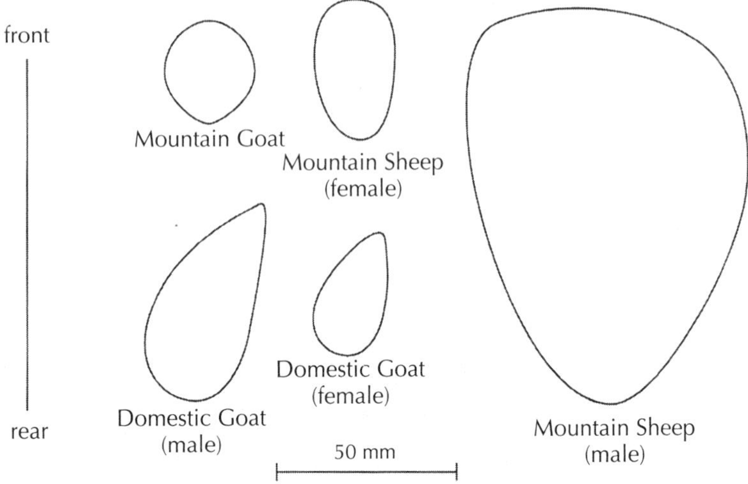

Figure 52. Basal cross-sections of left horn bone cores.

12a. Bone core < 100 mm long; horn sheath < 350 mm 13

12b. Bone core > 100 mm long; horn sheath > 350 mm 14

13a. Bone core sharply pointed and round in cross-section (figure 52); horn sheath is shiny black, with a smooth surface and sharp point, < 350 mm long and curved slightly back from the head; dorsal skull profile is narrow and tapers little from the cranium to the premaxillae (figure 53 A) .... Mountain Goat (*Oreamnos americanus*)

13b. Bone core blunt and oval in cross-section (figure 52); horn sheath is dull brown, with many rings and ripples along its surface, and has a blunt tip; dorsal skull profile is sharply tapered from the cranium to the premaxillae (figure 53 B) ... 16

14a. Bone core massive, with the base rounded to roughly triangular in cross-section (figure 52); horn sheath is curled in an open circle 15

14b. Bone core relatively slender, laterally compressed with a well-defined leading edge (figure 52); horn sheath extends upward from the skull in a gentle curve, or is flared sideways in an open spiral .... Domestic Goat (*Capra hircus*), horned breed

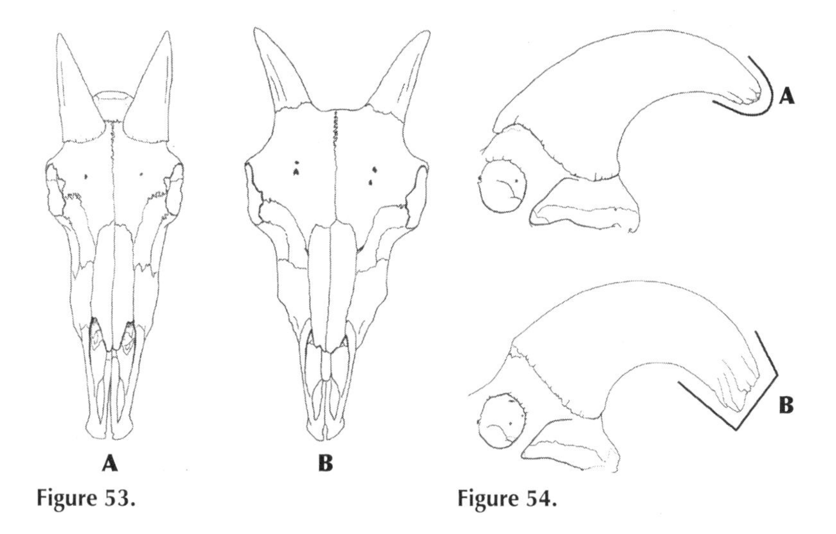

Figure 53.

Figure 54.

15a. Bone core massive, with a pointed tip (figure 54 A); horn sheath has a prominent keel on the outer edge
.... male Thinhorn Sheep (*Ovis dalli*)

15b. Bone core massive, with a blunt, jagged tip (figure 54 B); horn sheath lacks a prominent keel on the outer edge
.... male Bighorn Sheep (*Ovis canadensis*)

16a. Lateral profile of the skull over the frontals and nasals is flat (figure 55 A) .... female Bighorn Sheep (*Ovis canadensis*)
.... or female Thinhorn Sheep (*Ovis dalli*)

16b. Lateral profile of the skull over the frontals and nasals is convex (figure 55 B) .... Domestic Sheep (*Ovis aries*), horned breed

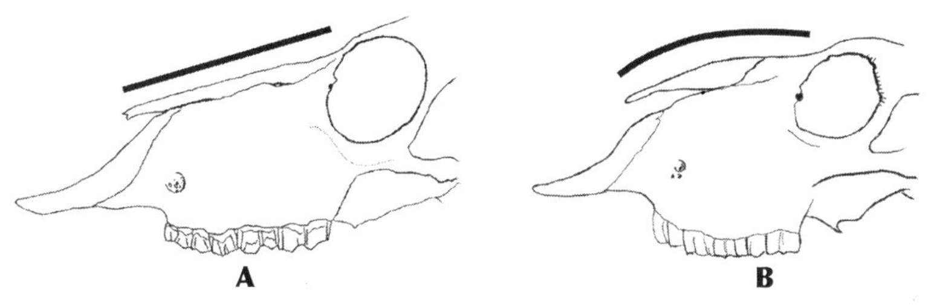

Figure 55.

17a. Basilar length > 350 mm
............................Domestic cattle (*Bos taurus*), hornless breed

17b. Basilar length < 350 mm 18

18a. Lateral profile of the skull over the frontals and nasals is convex (figure 55 B) ....Domestic Sheep (*Ovis aries*), hornless breed

18b. Lateral profile of the skull over the frontals and nasals is flat
.... Domestic Goat (*Capra hircus*), hornless breed

SPECIES ACCOUNTS

This book provides a detailed account for each of the province's nine species (including 17 subspecies) of native hoofed mammal and for its one introduced wild species. The accounts follow the same order as in the checklist. Each species account is divided into nine sections, and where appropriate, they include additional information for each subspecies. The nine sections are:

Other Common Names – A list of alternative English common names.

Description – A concise description of the species or subspecies. The descriptions for species with no subspecies list all quantitative data. Where there are subspecies, the overall species description includes only dental formulae, while the subspecies descriptions give body and skull measurements. I made every effort to use measurements from BC specimens (either live or museum specimens). But where there were none available, I used data on the same subspecies from other parts of western North America. Descriptions of body shape and coat coloration, as well as other characteristics, are based on my own observations of live animals and museum specimens together with published information.

Linear measurements are in millimetres (mm) and weights are in kilograms (kg). Where available, values are given for adults of each sex separately and provide the average, range and sample size (the number of specimens measured). For example, the hind-foot length of adult female Rocky Mountain Elk is given as "631 mm (600-670) n=15" to show that the average length of the hind foot is 631 mm, and that the average is based on measurements made on 15 specimens whose hind feet ranged from 600 to 670 mm in length. A question mark indicates that the relevant data were not reported in the information source.

Dental formulae represent the number of teeth in one side of the head (see figure 43). The first value is the number of teeth in the upper jaw and the second is the number in the lower jaw. For example, for Woodland Caribou, "incisors: 0/3, canines: 1/1, premolars: 3/3, molars: 3/3" indicates no upper but three lower incisors, one upper and one lower canine, three upper and three lower premolars, and three upper and three lower molars.

The identification information describes characteristics that separate the species from other ungulates that look most similar to it. Identification criteria to distinguish subspecies are given under the separate subspecies descriptions. Information is provided on external features as well as on the distinguishing features of the skull. Where useful, other identification information such as tracks and feces are described. Three views (dorsal, lateral and ventral) of the skull are given for each species along with a lateral view of the lower jaw. The scale line accompanying the skull drawings is 50 mm long. For all but Fallow Deer and Mountain Goat, adults of both sexes are shown in the drawings of live animals; in the case of Thinhorn Sheep, both subspecies are represented because they look distinctly different. For most other species, the external differences between subspecies are less obvious, except for the rumps and tails of Columbian Black-tailed Deer and Rocky Mountain Mule Deer. The rumps and tails of these two subspecies along with that of White-tailed Deer are shown for comparative purposes (figure 41).

Natural History – Information on the habitat, diet, social organization and grouping, social behaviour, reproductive data, criteria for determining age, life expectancy, and predators and other mortality factors. Information in this section was taken from studies of BC species and also from other studies in western Canada and the western United States.

Range – A general description of the overall distribution of each subspecies within British Columbia. This section also includes a brief history of any notable transplants and reintroductions. The accompanying range maps are based on maps originally developed by Dan Blower of the BC Wildlife Branch (Ministry of Environment) in conjunction with regional wildlife biologists. I updated these maps for this handbook with data supplied by regional wildlife section heads and their staffs. I have maintained the three relative density levels from the original Wildlife Branch maps to illustrate that animals are not found equally within a given distribution range. These densities are meant only to be approximate guides; ungulate numbers will vary within an area at different seasons, primarily because the animals move. Areas where there is a good chance of seeing a subspecies in BC are also included in the Range section.

Conservation Status – Information on the conservation status, along with a population estimate for each subspecies based on data from the BC Conservation Data Centre and the Wildlife Branch's 2011 estimates, respectively. If a species or subspecies is on the Red or Blue lists of animals at risk in BC this is stated but not if it is on the Yellow List.

Taxonomy – Information on related species and discussion of any relevant taxonomic issues.

Traditional Aboriginal Use – A brief summary of the major traditional value and uses of the species for First Peoples in the province.

Remarks – The origin of the species' name along with other information about the species in British Columbia that does not fit readily in other sections.

Selected References – A list of important publications, including those used in the account. It is not meant to be exhaustive, but should provide a starting point for anyone interested in additional information.

Elk

Cervus elaphus

Other Common Names: American Elk, Wapiti.

Description

The Elk is the second largest member of the deer family in British Columbia. It has a large body and long, slender legs. In winter, both sexes have a dark, sometimes blackish-brown head and neck, with a lighter greyish-brown body. The chest, legs and underbelly are usually darker than the rest of the body. There is a distinctive yellow-brown or cream-coloured rump patch, and similarly coloured, short tail. The rump patch is bordered on either side by a band of dark brown hair along the lower sides running up from the top of the dark legs. In winter, the longer hair on the neck, especially on the underside, forms a mane that is longer in males than females. The lower lip is light coloured with a distinct black stripe running down from the back of the mouth, and there is a ring of light hair around the eye. The summer pelage is generally the same colour pattern but somewhat more reddish-brown or tawny than in winter. Most adult males have a lighter coloured body than the females, with darker legs and neck – so a male has more contrast between the light and dark areas of his body than a female has, especially in early fall at the beginning of the rut. Antorbital glands are present and visible; other epithelial glands include metatarsal, caudal and interdigital glands.

Young up to about three months of age have reddish-brown coats with white spots. They lack the distinctive cream-coloured rump patch until they are about one year old.

Antlers are grown almost exclusively by males (females can grow antlers but only very rarely). They range in size and structure from a single spike or sometimes a simple fork in yearlings to the large branched antlers of adults. The antlers of adult Elk are characterized by a long, cylindrical main beam with usually five or six points branching from it. The first tines are two curved brow points protecting the eyes and head, followed by a third curved point midway along the beam, then a fourth large, relatively straight, upward-forward-pointing tine three-quarters along the beam. After the fourth tine, the main beam bends backward and ends usually either in a fork or a single sharp point. Large specimens can have one or two more tines after the fourth point. The same animal can have a different number of points on each antler. The average length of the main beam from the base

to the tip for 67 adult male Rocky Mountain Elk from Alberta was 1,186 mm, with a maximum length of 1,380 mm. The term "spike" or "spikehorn" is sometimes used for yearling males because of their simple antler structure, while "raghorn" is sometimes used for two- to three-year-old males whose slender three- to five-pointed antlers are easily damaged. Antlers of older males are stronger and heavier, and each usually weighs over 2.5 kg. Elk shed their antlers beginning in late February for the largest males and extending to late April and even early May for younger ones. New antler growth begins soon after shedding.

The skull of an Elk is large, with well developed lachrymal pits, open lachrymal vacuities, posterior nares undivided by the vomer and a broad premaxillae. It is one of only two species of deer in British Columbia with upper canine teeth; these short, wide and rounded teeth are present in both sexes (figure 47).

Measurements:
See subspecies descriptions.

Dental Formula:

incisors: 0/3
canines: 1/1
premolars: 3/3
molars: 3/3

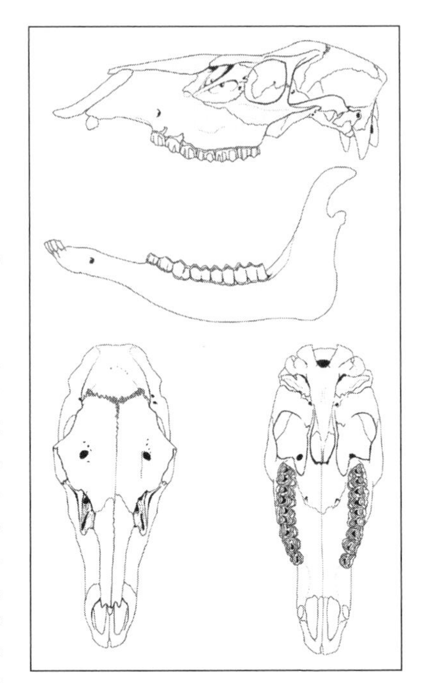

Identification:

Being the second largest deer in British Columbia, the Elk is much larger than the Woodland Caribou, White-tailed Deer and Mule Deer. The Moose is the closest in size, but it has a dark body with lighter coloured legs, and a long head and bulbous nose. In contrast, the Elk has a fawn or reddish-brown body with darker legs and neck, a compact, dark hairless nose, and a large light-coloured rump patch. The Elk's antlers also differ markedly from the palmated antlers of the Moose. At maturity, Elk antlers typically have a long main beam growing up and back from the head, with up to six long, pointed tines on each beam.

The next most similar species to Elk is Woodland Caribou, but again, coloration and antlers differ. The Elk is reddish brown, not dark greyish or chocolate brown as is the Caribou, and has a dark neck compared to the light neck of the Caribou. The Elk also lacks the white ring of hair above the hooves and does not make a clicking sound when walking. Finally, female Caribou often have small antlers while female Elk almost never grow them, and antlers of adult male Caribou have palmated tines with rounded ends, unlike the sharp pointed ones of Elk.

In skulls, the presence of upper canine teeth indicates either an Elk or a Caribou. But Elk canines are much larger, have rounded ends and project further below the premaxillae (and gum) than those of Caribou. Unlike Caribou, the premaxilla in the Elk skull contacts the nasal bone; when viewed from above, the shape is rounded, lacking the angular profile of Caribou (figure 47). Finally, in an Elk skull the vomer does not divide the posterior nares (figure 47) as it does in the Caribou skull.

Elk tracks are slightly smaller but less pointed at the front than those of Moose and are proportionally narrower than those of Caribou, Cattle or Bison (figure 38). Except for the soft spring feces, dry adult Elk pellets are usually dimpled at one end and have a small projection at the other, giving them an almost acorn shape. Pellets of Odocoilid deer are about half the size and lack these end characteristics. Moose pellets are about the same size as those of Elk but they are usually rounded at either end.

Natural History

Rocky Mountain Elk prefer feeding in a mix of open grasslands and shrub lands or in open, mixed conifer and deciduous forests; they also prefer forested habitats for resting. Like Roosevelt Elk, Rocky Mountain Elk will also forage in avalanche paths in summer. Roosevelt Elk, inhabiting the southern coastal rainforests, are usually found in valley bottoms in most seasons, even in summer in some areas. Herbs and stands of shrub seedlings, along with riparian areas, provide the main foraging areas for Roosevelt Elk, while older forests supply security cover against predators. Both subspecies seem to prefer bedding just inside the forest edge where they can remain hidden yet retain a clear view to watch for predators. In winter when snow depths exceed 30 cm, Roosevelt Elk move into the mature and old-growth forests where snow is not as deep because of the dense tree canopy above (see colour photograph C-10). In cold, low-snow periods, they may move onto open south-facing slopes where the sun's warmth is strongest, and if necessary, use young, mature or old-growth stands that provide them with thermal cover.

Seasonal movements can be highly individualistic and there may be no general pattern even within a population. Some individuals make seasonal altitudinal migrations as high as alpine grasslands in summer, some move mainly horizontally and others remain year-round in the same general area. Seasonal use of space seems to be influenced by various factors including local vegetation, snow conditions, predators and past experience.

Elk are generalist herbivores with a mixed browser-grazer diet and a high degree of dietary flexibility. Consequently, Elk tend to show diet overlap with many other species of ungulates. The relative importance of forages varies with subspecies, season and even the herd. Rocky Mountain Elk consume a variety of grasses, sedges and forbs, as well as shrubs such as willows, Soapberry, Saskatoon and Trembling Aspen. Their winter diets are primarily grasses, sedges and shrubs, with forbs becoming important in spring and summer. Rocky Mountain

Elk probably eat more grasses, sedges and forbs than Roosevelt Elk do, and use more open grass-shrub communities. Roosevelt Elk, on the other hand, live mainly in more heavily forested habitats, which is reflected in their diet. Important browse species for Roosevelt Elk include Red Elderberry, various *Vaccinium* and *Rubus* species, willows, Dull-leaved Oregon-grape, Pacific Ninebark, Amabilis Fir and Western Hemlock. These Elk can have major impacts on newly planted tree seedlings such as Western Red-cedar. Roosevelt Elk also eat various grasses and sedges, and many types of forbs, such as Skunk-cabbage, in spring and summer. Sword Ferns and Deer Ferns are also important forages in most seasons except summer. Preferred forages used by both subspecies grow in early successional stages following a fire, or in young cut-overs following logging. Riparian areas are also important for foraging, especially for Roosevelt Elk. When not hunted, Elk adapt well to humans and find lawns and golf courses excellent places to graze.

Elk are highly social animals that congregate in large maternal groups. But, as with most hoofed mammals in BC, the mature males and females live apart for most of the year. Males leave their maternal groups when 2.5 to 3 years old and socialize with females only during the mating season.

Maternal groups can be quite large, and their size depends in part on habitat; larger groups tend to develop in more open areas. Adult female and juvenile Rocky Mountain Elk form the largest groups, but on Vancouver Island, maternal groups of Roosevelt Elk can number more than 100. Such large groups are not necessarily stable and may separate into smaller units. All-male groups of similarly aged individuals are much smaller and may be even less stable over time. While feeding or walking in groups, Elk often make a low grunting call, perhaps to maintain contact among members.

The mating or rutting period usually begins in September and extends into October. Males are especially active. Their best known and most obvious behaviour is perhaps their bugling vocalization. This hollow squealing whistle starts with a two-note call, extends into a long higher note that drops suddenly in pitch, and usually ends with two or more separate short whistles or grunts. While bugling, the male stands with his mouth open; the last short calls are usually accompanied by sharp contractions of the belly, and often by spurts of urine. These calls appear to serve as challenges to other males and warnings to stay away. Male Elk also threaten each other by approaching an opponent with head and nose raised and the upper lip curled to reveal the small canine teeth, while making a hissing sound. Besides

Figure 56. Similar sized males engage in serious fights during the height of the rut, when competing for females. Before then, sparring seems to allow them to practice their fighting skills and to learn about the abilities of their opponents.

the bugling and head-up threat, an adult male will make noises and threats by thrashing bushes and rubbing small trees with his antlers, leaving vertical gouges edged by frayed bark. A male Elk will also paw the ground and urinate on it, then wallow on the wet ground. Soaking his mane and body with urine increases his body odour, which probably has signal value to other males as well as to females. (Wet wallows made during the rut are different from the drier ones used in summer, which Elk use to discourage biting insects. Elk may rub their antlers against small trees and saplings at other times of the year as well, and signs of this activity can be concentrated in some areas.)

Even before the rut begins in earnest, males will spar and fight with each other. Large males of similar size often trot or walk beside each other a few metres apart for up to 200 metres, before turning and repeating the parallel walk. During these walks, they may tip their antlers toward each other, which sometimes leads to a fight. They fight head to head after engaging their antlers, pushing and twisting as they try to throw each other off balance so they can try to gore their opponent with their antler points. When fighting head to head, the long brow tine helps protect the face and eyes from damage, but antler points can break. Serious fights between equally matched males can sometimes lead to injury and, occasionally, to death. Females – and males without hard antlers – fight by rearing up on their hind legs, ears back, flaying their front legs rapidly as they strike at their opponent with the front hooves. Sometimes just holding the head high with ears back is a sufficient threat to end a dispute.

Elk are polygamous and adult males defend a harem. Most harems consist of one adult male and up to 20 (or more) adult females and young. The harem male excludes all other males except calves. He must constantly defend the females from the satellite males hanging around the outer edges, trying to steal females or quickly copulate with them. He chases away rivals by bugling and using the threats described above. The harem male has little control over the movements of the females. He can only try to keep the group together by using a specific herding posture. He circles a female from her rear or side in a broadside display, head low but with the nose tipped up, so the antlers are alongside his neck. If the female does not move back into the group, the male usually rushes at her, bringing his nose down rapidly so the antlers swing forward and down to push or bump her rump.

Courtship patterns are not particularly elaborate in Elk. The male's courtship approach is the opposite of the herding posture. The courting male Elk moves toward the female from the front with his head and antlers raised slightly, flicking his tongue (figure 27) and making licking or soft grunting sounds. The female often responds by lowering her head, neck extended, sometimes ears back, while making biting movements, opening and closing her mouth rapidly (figure 25). The courting male sniffs her rear, lip-curls, and puts his neck over her rump. Like most deer, copulation takes place during a brief copulatory jump, in which the female stands still and the male grasps her with his forelegs just in front of her pelvis. He makes a single thrusting ejaculatory jump with the head thrown back and the hind feet often leaving the ground momentarily.

Gestation lasts about 255 days. Just before giving birth, females often leave the herd to have their young, sometimes staying secluded for a time ranging from a few days to about two weeks. Females usually bear one young weighing 10 to 15 kg, in late May or early June. Twins are rare. For the first few weeks of life, the young are hiders (see page 49), usually resting along the forest edge or in other vegetation cover, while the remainder of the herd forages nearby. When in danger, a young Elk makes a high screaming sound to bring its mother running to its rescue.

Most females give birth to their first young in their third year, but in some populations it may be a year earlier. Males can produce sperm at 1.5 to 2 years, but usually do not become fully sexually active and take part in the rut until at least 3 years old, and for most, not until they are about 7 or 8. An Elk's age can be determined by tooth succession and by cementum annuli counts using the first incisor or first lower molar; tooth wear is not always a reliable method. Elk can be long-lived, but the average longevity varies between populations. Females usually live longer than males, reaching 20 to 24 years compared to 12 to 14 years for males.

Grey Wolves are the main predators of all age-sex classes of Elk. Cougars also prey on Elk, and both Black and Grizzly bears prey on young of the year. A nematode worm, *Dictyocaulus* sp., may be a significant parasite for Rocky Mountain Elk, especially during periods of poor nutrition. Domestic Cattle is the primary host for this parasite, although it is also reported in Odocoilid deer. The parasite lives in the air passages; when present in large numbers, it can make breathing difficult. This can lead to respiratory infection, with or without other disease organisms, and result in bronchitis. The Winter Tick can be a serious parasite of Elk, especially when severe winters and shortage of winter range contributes to malnutrition. The large American Liver Fluke is common in populations of both subspecies of Elk in BC; it seems to have little effect on Elk, but it can kill Moose and domestic ungulates. The life cycle of the Liver Fluke involves aquatic snails, so the fluke is rarely seen in regions with predominantly dry habitats. The liver of any infected animal may have large cavities and tracks filled with immature flukes, surrounded by quantities of soft, blackish material. Throughout the range of Elk, severe winters with high snowfall can lead to malnutrition and increased mortality either from starvation or increased susceptibility to predation.

Rocky Mountain Elk

Cervus elaphus nelsoni

Other Common Names: Wapiti.

Description

The Rocky Mountain subspecies differs little from the species.

Measurements:

weight -
- male: 353.5 (304-401) n=10
- female: 245.5 (168-311) n=38

total length -
- male: 2,410 mm (2,310-2,510) n=11
- female: 2,274 mm (2,080-2,480) n=15

tail vertebrae -
- male: 137 mm (102-160) n=11
- female: 143 mm (100-180) n=15

hind foot length -
- male: 672 mm (640-690) n=11
- female: 631 mm (600-670) n=15

shoulder height -
- male: 1,295 mm n=1
- female: 1,346 mm (1,270-1,422) n=4

ear length -
- male: 211 mm (190-230) n=11
- female: 203 mm (180-220) n=15

skull length -
- male: 466.0 mm (448-485) n=3
- female: 431.7 mm (412-454) n=29

skull width -
- male: 186.3 mm (170-195) n=3
- female: 170.8 mm (160-180) n=34

Range

The distribution area of Rocky Mountain Elk in British Columbia has increased significantly since 1965, when Cowan and Guiguet reported it in *The Mammals of British Columbia*. Today, the main distribution areas in the province are in the south-central, the southeast and north-east regions. The most extensive of these three areas is in the northeast in the Peace River region. There, it extends as far north as the Liard

Plateau, west along the Liard River valley and then south along the Kechika River valley and surrounding tributaries, as far west as the headwaters of the Major Hart and Turnagain rivers, and as far south as Mount New. Most of the Elk population in the Peace-Liard region is sparse, but there are some areas of higher density, the most southerly of which is along the Murray River. There are also two small areas of high density to the west of Fort St. John, and also further north in a larger area that extends south from the Tea River to the Prophet River.

The next largest distribution area, and the one with most of the province's Elk, is in the southeast. This region lies between the Columbia and Rocky mountains and extends from Rogers Pass and Donald Station south to the international border. Two other areas with moderate densities of Elk are the east side of the southern Okanagan Valley and just northwest of Princeton along Granite, Tulameen and Allison creeks. Elsewhere, Elk occur in low densities and sometimes in small isolated pockets, such as along the Stuart River west of Prince George, in the Robson Valley east of Prince George, around the Williston Reservoir and along the Bella Coola River upstream of the town of Bella Coola. Isolated sightings of Elk, usually males, are occasionally made on the Chilcotin Plateau and around the junction of the Parsnip and Anzac rivers, but none of these seem to represent any locally established population. Rocky Mountain Elk living between the Chilliwack River and the international border are the result of an expansion of a population that was introduced into northwestern Washington.

The present distribution of Rocky Mountain Elk is thought to be about the same as it was in the late 1700s and early 1800s, before the major decline in the mid to late 1800s. It is uncertain why Elk numbers declined, but starting in 1917, the provincial government made at least 21 reintroductions to increase Elk populations throughout BC. This program, along with other measures, was generally successful. Rocky Mountain Elk were reintroduced principally from the East Kootenay and Alberta, to many areas in central, north-central and southern BC, including Adams Lake north of Chase, Lillooet, Ingenika River, Dunlevy Lake, Naramata, Lytton and east of Christina Lake. Rocky Mountain Elk were also introduced onto Graham Island, Haida Gwaii, in 1929, using animals from Buffalo Park near Wainwright, Alberta, and in 1930, with Elk from the Okanagan that had originated from Wainwright. At one time, Elk were on Moresby Island. They still exist on Graham Island, where they may have interbred with the European Red Deer that were introduced there in 1918. Today, there are perhaps 100 to 150 Elk on Graham Island. For other introductions see Remarks on page 108.

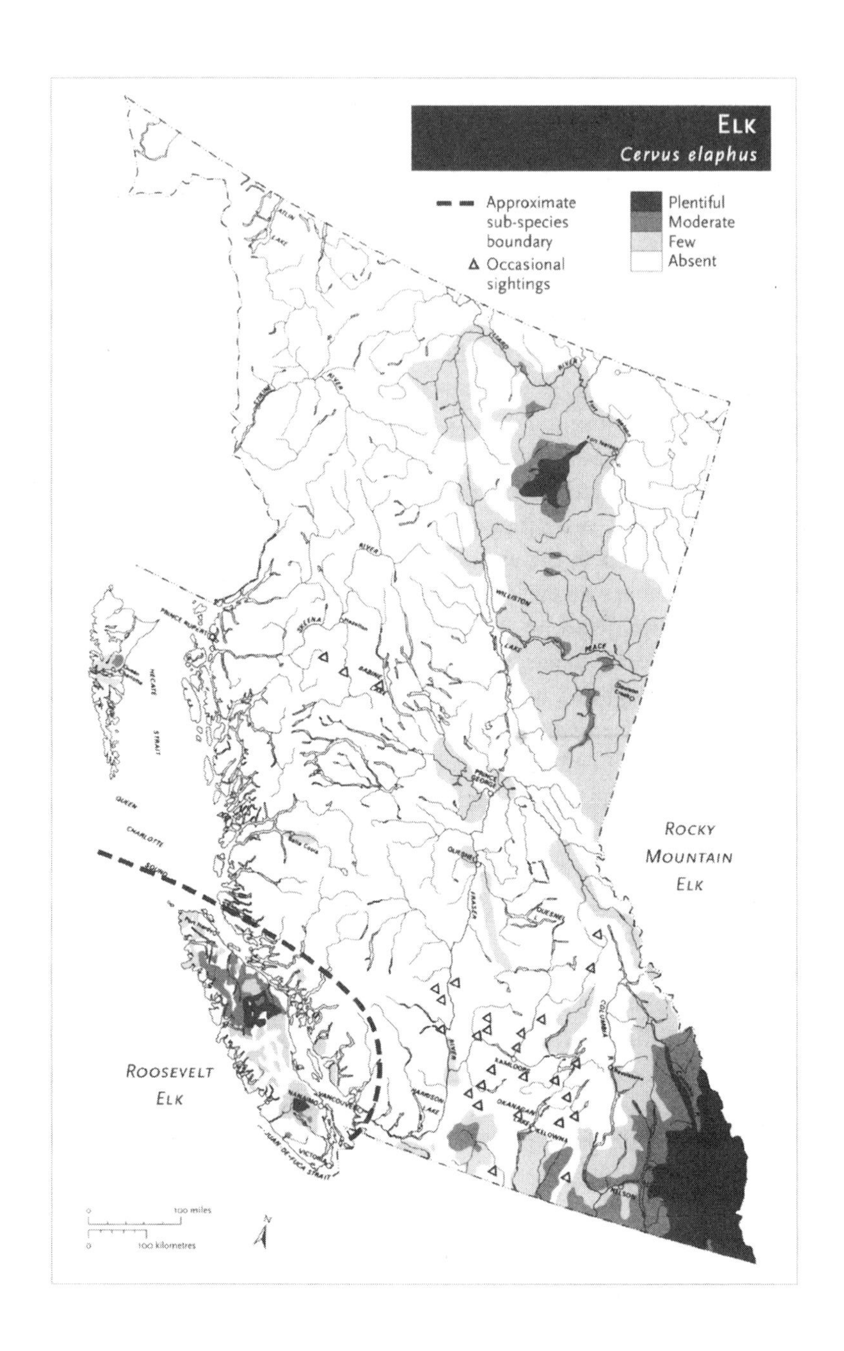
ELK
Cervus elaphus
Approximate sub-species boundary
Occasional sightings
Plentiful
Moderate
Few
Absent
ROCKY MOUNTAIN ELK
ROOSEVELT ELK
100 miles
100 kilometres

Recently, Rocky Mountain Elk have increased in the East Kootenay and Peace-Liard regions. In the 1970s and 1980s, their numbers increased in the East Kootenay, but began to decline in the 1990s due to factors such as range deterioration, predation, severe winters in 1995-96 and 1996-97, and liberal hunting limits set to reduce the population. Deterioration of the Elk's range was caused by factors such as forest encroachment, competition with livestock and habitat loss due to land developments (e.g., hobby farms, golf courses, recreational property). In the Peace-Liard region, Elk numbers have at least tripled since the 1970s, after having all but disappeared between the 1850s to the late 1920s. This increase in the northeast was probably partly a response to extensive, widespread prescribed burning that created favourable habitat for Elk. In the Chilcotin region of central BC, Elk populations have disappeared, but it is evident from dating antlers found in bogs and the bottom of ponds, that the species inhabited the region 150 or more years ago.

Most Rocky Mountain Elk are found in the Kootenays (50%), and the Peace-Liard region (45%). Some of the best viewing opportunities include Kootenay National Park during spring and fall, around Field in Yoho National Park year round and at various locations in the East Kootenay Trench south from Golden in winter and spring.

Conservation Status

With between 38,000 and 72,000 Rocky Mountain Elk estimated to live in the province in 2011, and most populations stable, this subspecies is not threatened in BC, although some regions report declines.

Roosevelt Elk

Cervus elaphus roosevelti

Other Common Names: Vancouver Island Elk, Olympic Elk.

Description

Roosevelt Elk is the largest subspecies of Elk in North America. Its body colour is often darker than that of Rocky Mountain Elk, and the male's neck can sometimes be almost black in colour. The neck of adult female Roosevelt Elk is also dark brown. In summer coat, the main body colour of both sexes is reddish brown. The antlers of adult males are less backward oriented, usually shorter and more robust than those of Rocky Mountain Elk. In older males, the antlers typically terminate in a crown, formed by a cluster of three or more points beyond the fourth tine; some minor palmation or webbing can develop in this area as well. Such a crown is similar to that found in European Red Deer; I have also seen it occasionally in large antlers of Rocky Mountain Elk. See also the description for the species on page 94.

Measurements:
weight -
  male: 303 kg n=1
  female: 261 kg (228-298) n=9
total length -
  male: 2,267 mm (2,172-2,362) n=2
  female: 2,231 mm (2,127-2,331) n=9
hind foot length -
  male: 658 mm (616-690) n=2
  female: 655 mm (629-673) n=9
shoulder height -
  male: 1,518 mm (1486-1549) n=2
  female: 1,446 mm (1,359-1,543) n=9
skull length -
  male: 435.5 mm (426-446) n=4
  female: 406.9 mm (396-419) n=7
skull width -
  male: 189.8 mm (185-192) n=4
  female: 176.3 mm (171-179) n=7

Range

Most Roosevelt Elk in British Columbia live on Vancouver Island where they arrived at least 3,000 years ago. They are found over nearly all of the island except at the southern end and along most of the west coast south of the Brooks Peninsula. There are also small pockets on the mainland including the head of Phillips Arm and Loughborough Inlet, around Powell River, and on the Sechelt Peninsula. The animals near Powell River and the Sechelt Peninsula are the result of reintroductions mainly from Vancouver Island: the two to the Powell River area were made in 1994 and 1996, and the three to the Sechelt Peninsula were carried out between 1987 and 1989. Small numbers of this subspecies are now being reintroduced to areas in the Lower

Mainland. South of BC, Roosevelt Elk extend west of the Cascade Mountains, south through western Washington and Oregon as far as northern California. In southern California, Roosevelt Elk are replaced by the Tule Elk.

Although Roosevelt Elk inhabit areas with dense vegetation, there are several places on Vancouver Island where there is a good chance of seeing them. These include the Elk River valley in Strathcona Provincial Park just west of Buttle Lake, Green Mountain, Shaw Creek and the White River. They can often be seen around Powell River near the local golf course. In the rutting season, you are likely to hear Elk before you see them. Bugling males can be heard over quite long distances. You can buy or make an Elk caller to attract rutting males. But use it with caution, because a caller can be effective, and it can be quite startling to be confronted by a surly, rutting male Elk only a few metres away, appearing seemingly from nowhere without a sound.

Conservation Status

In 2011, the BC population of Roosevelt Elk was between 5,900 and 7,100, almost all (4,600 to 5,600) on Vancouver Island. The densest populations were in the Salmon, White, Adam, Eve, Tsitika and Nimpkish river drainages. Although the total number of Roosevelt Elk is relatively low in BC, careful management has so far ensured their survival. But poaching is a growing problem, because these Elk live mainly at low elevations where they are easily found by poachers driving logging roads.

Taxonomy

Elk are part of the Red Deer group that is currently considered to be a Holarctic (across the northern hemisphere) super-species of *Cervus elaphus*. Members of this group are found from the United Kingdom, across Europe, and scattered subspecific populations throughout Central Asia, northern and western China, southern Siberia, and into North America as far south as the Mexican border.

Skull dimensions have been used to separate the subspecies of Elk, but are probably of limited value compared to external characteristics. Differences in skull size between the island and mainland populations of the same subspecies in Washington and Oregon may be the result of interbreeding with Rocky Mountain Elk on the mainland.

Small numbers of the latter subspecies were introduced periodically into parts of central Washington and Oregon in the days before wildlife managers recognized the dangers of such actions. The type locality for Rocky Mountain Elk is Yellowstone National Park, Wyoming and for Roosevelt Elk is Mount Elaine on the Olympic Peninsula, Washington.

Traditional Aboriginal Use

Interior and northern aboriginal groups hunted Rocky Mountain Elk extensively, and some southern Coastal tribes hunted Roosevelt Elk. They used the smaller Elk bones, such as the metapodials, to make hide scrapers, chisels, adzes, needles and awl containers, and the larger bones to make splitting wedges. These groups also made knife blades, pendants and animal callers from Elk bones, and used the bone marrow as food. Many groups used Elk bone and antler to make barbed harpoon heads, blanket pins and pestles for grinding pigments. They also fashioned antlers into clubs, digging tools, tool handles, combs and ornaments. The peoples on the southern Coast perhaps valued them most as wood-splitting wedges and adze handles.

Elk hides were a source of clothing articles such as leggings, moccasins, dance aprons, pouches and bags. A sleeveless shirt or battle armour could be made by sewing and gluing two or three layers of Elk hide together. Aboriginal peoples also used the skin to make balls for games and the rawhide for sewing thread. In the early 19th century, aboriginal groups in northern BC obtained Elk hides from Hudson's Bay Company traders who had acquired them on the southern coast. People in the southern interior made pendants from the upper canine teeth of Elk and sometimes sewed them along the fringes of a woman's shirt. Today, Elk canines (ivories) are sought by many non-aboriginal people in North America.

Remarks

Cervus is Latin for "deer" and *elaphus* is Greek for "deer". The name "elk" comes from the German *elch* which is the common name for European Moose, and was misapplied by early European immigrants in North America. "Wapiti" is the name for Rocky Mountain Elk in the Shawnee language and means "white rump". Some people today prefer "Wapiti" over "Elk" as the common name for the species. Roosevelt Elk was named after U.S. President Theodore Roosevelt.

European Red Deer, originating from Scotland, but imported by way of New Zealand, were introduced onto Graham Island, Haida

Gwaii, in 1918. Apparently they survived until the 1940s, but there have been no recent sightings since, so they are presumed to be eradicated. But European Red Deer may have interbred with Rocky Mountain Elk that were also introduced to Graham Island around 1928.

Besides the numerous introductions and reintroductions of the native subspecies of Elk in the province, there were also two introductions of Manitoba Elk from Elk Island National Park, Alberta: the first transferred 24 animals into the Lardeau River valley at the north end of Kootenay Lake in 1949; and the second brought 57 animals into the Kechika River valley in 1984. The following year, 68 Rocky Mountain Elk were released in the same area from the East Kootenay. These two subspecies have no doubt interbred in both areas, underscoring the importance of avoiding transplants of non-native subspecies. Another problem of introducing animals is the spread of diseases and parasites. The Giant Liver Fluke found in some Elk in the East Kootenay can be passed on to Moose and other ungulates, so care is obviously needed in transplanting animals to new areas. Readers interested in further details should consult David Spalding's 1992 review. This provides a wealth of interesting detail about the past and present distribution and numbers of Elk throughout the province, along with information on the many transplants that have been made.

In the 1980s, the large Rocky Mountain Elk population in the East Kootenay created major conflicts with ranchers, because the Elk were having a significant impact on grasslands and especially on haystacks. Current proposed management includes reducing Elk herds in some key areas.

Care needs to be taken when around adult male Rocky Mountain Elk during the mating season, particularly in areas where they are accustomed to people, such as national parks.

Selected References: Blood and Smith 1984, Brunt 1990, Leslie and Jenkins 1985, Nyberg and Janz 1990, Quayle and Brunt 2003, Schonewald-Cox et al. 1985, Spalding 1992, Vore and Schmidt 2001, Wisdom and Cook 2000.

European Fallow Deer
Dama dama dama

Other Common Names: Fallow Deer, Fallow.

Description

The European Fallow Deer is an average-sized deer with slender legs and medium-sized ears. The males have a conspicuous bump (the larynx) on the throat, like an Adam's apple. There are several colour variations. The summer coat colour in the most typical form is a light reddish brown with white spots scattered over the back and sides, and a horizontal row of spots that sometimes merge into a single stripe midway along each side, often with a darkish horizontal stripe below. Many individuals with this coat type have a narrow, dark dorsal stripe running from the nape of the neck to the tail. The European Fallow Deer also has long individual hairs scattered over its body. The belly is white and the undersides of the neck and chest are also light or white. The tail is moderately long and covered with long hair from the tip to the base; it is usually dark brown or black on the upper surface and white on the under surface. Except for a patch of bare skin around the ano-genital region, the hairs on the rump beneath and on either side of the tail are also white with a contrasting outer black stripe on either side that may connect at the base of the tail. These two stripes create a heart-shaped pattern around the tail. In winter, the upper body becomes a dark reddish brown and the spots may be difficult to see. Males often have a small dark cap from the base of the antlers to just between the eyes. Because this dark cap stops just above the eyes, there can be the appearance of a light upper eye-ring. The penis sheath ends with a conspicuous tuft of hair (the brush). Young of the year are spotted like the adults. Fallow Deer have antorbital, preputial (at the tip of the penis), metatarsal and rear interdigital glands. Females have four teats.

The European Fallow Deer has two other main colour phases: a non-spotted, creamy or buff white variety that lacks the dark cap, and a dark brown, almost black form that may or may not have faintly visible spots. Some of the dark forms without spots often lack the light belly, although in others it is a lighter brown. The dark brown forms usually retain the black tail and black marks on the rump. The young of these other colour forms are similar to the adult's coloration, although in the darker colour phases, light spots may be present.

The adult male Fallow Deer's antlers are erect and strongly palmate on the upper half. The main beam is round in cross-section for the first half of its length. Branching off this beam is a large, curved brow tine followed by another large, curved, forward-pointed tine about halfway along. The last third or half of the main beam ends in a palmated area with a series of small points projecting from its upper and rear edges, sometimes with one larger tine at its base that projects backward. The orientation of the antlers is such that they sweep back and up from the head, with the palmation visible in greatest profile from the side.

Measurements:

weight -
- male: 57.0 kg (42.2-76.2) n= 15
- female: 36.6 kg (33.8-43.1) n=13

total length -
- male: 1,537 mm (1,400-1,640) n=15
- female: 1,365 mm (1,230-1,450) n=13

tail vertebrae -
- male: 178 mm (160-200) n=11
- female: 158 mm (130-200) n=13

hind foot length -
- male: 411 mm (380-420) n=15
- female: 376 mm (360-400) n=13

shoulder height -
- male: 856 mm (760-890) n=15
- female: 755 mm (700-830) n=13

chest girth -
- male: 860 mm (700-1,000) n=15
- female: 717 mm (660-780) n=13

skull length -
- male: 221.7 mm (201-248) n=22
- female: 208.8 mm (202-221) n=9

skull width -
- male: 103.7 mm (97-118) n=21
- female: 96.4 mm (94-102) n=10

Dental Formula:

incisors: 0/3
canines: 0/1
premolars: 3/3
molars: 3/3

Identification:

In adults, the coat colour of all colour variants differs from that of any of the Mule Deer group or the White-tailed Deer. Fallow Deer are the only deer in British Columbia in which the adults have a spotted coat. The small white rump patch with its heart-shaped black border is also distinctive of Fallow Deer. The antler form with its main beam and first two forward pointing tines, followed by the palmated upper half, wider at the top than at the bottom, is unique among any deer in BC.

For skulls, the vomer does not divide the posterior nares in Fallow Deer, and the shape of the opening on its ventral side is more pointed than it is in Odocoilid deer skulls (figure 49). In the lower jaw, the first incisor is at least twice as wide as the others, and the premolars and molars, especially on the lower jaw, are larger than those in comparable-sized Odocoilid deer skulls. Tracks and feces would be difficult to distinguish from native Odocoilid deer such as Columbian Black-tailed Deer.

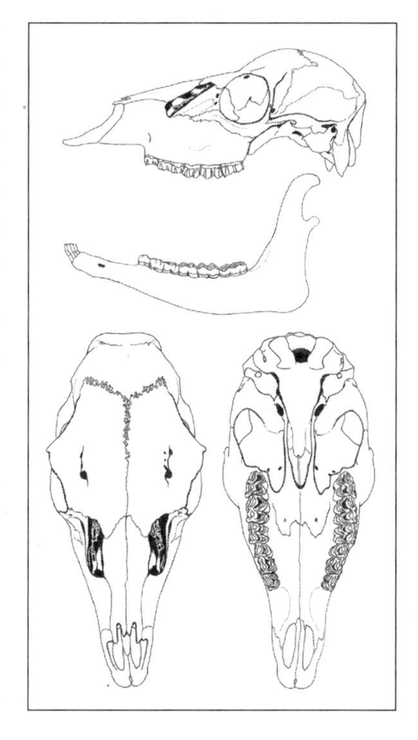

Natural History

European Fallow Deer graze and browse. They prefer open forests and adjacent grasslands that together provide both grazing and browsing opportunities. On the Gulf Islands, they will occasionally eat Eel-grass washed up on the beach. Evidence of heavy browsing on the islands is shown by a distinct browse line on the trees, and Fallow Deer will even stand on their hind legs to reach higher vegetation. They are active mainly at night between dusk and dawn when they use open areas to graze, but can be seen throughout the day in the forest. For most of the year, adult males live in small groups (usually two to five members), separate from females and young (who usually form groups of five to twenty).

On Sidney Island, the rut occurs in October, and probably peaks about mid October. Although Fallow Deer mating systems in British Columbia have not been studied, elsewhere they can vary among populations, perhaps in response to deer density. In some populations, a single male guards a group of females (his harem) against other males, while in others a behaviour called lekking occurs. Lekking is rare in mammals, although not uncommon in birds. Fallow Deer is the only deer known to engage in lekking behaviour. It involves males congregating in a relatively small area called an arena, on which each adult male establishes a small territory, only a few metres in diameter (colour photograph C-20). Territories are usually occupied around dawn and dusk each day when a male defends it from neighbouring males. He also displays to attract females onto his territory. When he succeeds, he tries to keep the females there long enough to court and copulate with them. Lekking is an amazing phenomenon to watch; the territorial males are almost constantly vocalizing and jerking their heads up and down. The call is a grunting snort, and the head nodding draws attention to their palmate antlers. While vocalizing, a territorial male also uses the tuft of hair on the end of the penis sheath to scatter urine over himself, probably along with glandular secretions from the preputial gland. Presumably, these odours attract females. When not trying to keep females on their small territories, males fight with neighbours, and smaller males rush in and try to steal females.

Gestation is believed to be between 150 and 220 days. Usually one young is born weighing between 2 and 4 kg; twins are rare on Sidney Island. Births in BC occur in late May and in June. The female gives birth in seclusion or occasionally within the herd; the young are hiders. Fallow Deer reach sexual maturity at 18 months. Females give birth for the first time either at two or three years of age. As is typical in most ungulates, most young males, though capable of breeding, are prevented from doing so by older, dominant males. Consequently, most male Fallow Deer begin mating when around five or six years old. But when overhunting removed many mature males on Sidney Island, younger males mated with females. The longevity of Fallow Deer in BC is unknown.

Fallow Deer have no major predators (other than humans) on the Gulf Islands that they inhabit. Old age and starvation are the most probable natural mortality factors. When alarmed, they may make a short barking sound and hold their tail up and curled slightly over the back showing the white underside. Alarmed Fallow Deer also use a less pronounced form of the peculiar bouncing gait called stotting that is more typical of Mule Deer. Fallow Deer are hunted on Sidney Island in a private section of the Island, and there is also a carefully regulated limited-entry hunt in the Sidney Spit Marine Park. At one time the population on James Island was periodically culled.

Range

Fallow Deer is an introduced species whose natural distribution is the Mediterranean region of the Middle East and southeastern Europe. Since at least the Neolithic, Fallow Deer have been widely introduced into many countries. In British Columbia, Fallow Deer were recently introduced from the Duke of Devonshire's Estate in England to James Island, probably around 1908. From there they apparently swam across to Sidney Island and by the 1950s were becoming well established. Reliable reports indicate their presence on James and Sidney islands, as well as Mayne, Samuel and Saturna islands. The Fallow Deer on Mayne are reportedly escapees from deer farms on the island and are probably the source of these deer on Saturna. In the early 1930s, people tried to introduce Fallow Deer from James Island onto two other southern Gulf Islands – Saltspring and Pender – as well as to the Alberni district on Vancouver Island, but fortunately, these attempts were unsuccessful. No recent sightings have been reported on Saltspring or D'Arcy islands, and although these deer were reported on the soutthern end of Galiano Island, they appear to have been removed. Fallow Deer have been seen in parts of the lower Fraser River valley where they are believed to have escaped from the Vancouver Game Farm. But as yet, there are no reports of them breeding on the mainland.

Outside game farms and zoos, Sidney Spit Provincial Park on the north end of Sidney Island probably provides the best opportunity to see these deer in BC, especially after the rut when many mature males move into the park. Boaters often see these deer along the shoreline.

Conservation Status

Permanent populations of Fallow Deer are known to inhabit only James, Mayne and Sidney islands in the Juan de Fuca Strait off the southeast coast of southern Vancouver Island. The population on Sidney Island is estimated to be over 600, while that on James Island is likely several hundred. In 1998, the population on Mayne Island was estimated to be around 30. There are no estimates for numbers of this species.

Taxonomy

Authorities disagree whether there are one or two species of European Fallow Deer, but most place it in its own genus, *Dama*, although some include it in the genus *Cervus*. If the two forms are considered separate species, the correct names are the European Fallow Deer (*Dama dama*) and the Mesopotamian or Persian Fallow Deer (*Dama mesopotamica*). If they are regarded as a single species, it is referred to as Fallow Deer (*Dama dama*) with two subspecies: *D. d. dama* and *D. d. mesopotamica*. Fallow Deer originated in the eastern Mediterranean region of southern Europe and Iran, and possibly parts of North Africa. The type specimen of European Fallow Deer came from an introduced population in Sweden.

Traditional Aboriginal Use

Being a recently introduced species, there were no traditional uses.

Remarks

Fallow Deer are a popular species for game ranching in the province. They are farmed for their meat (venison) and to a lesser extent for their antlers (velvet), which are not regarded as the best grade for the medicinal trade. On some areas of the southern Gulf Islands their feeding has led to severe negative impacts on native plants – in some cases this is exacerbated by introduced European Rabbits.

Selected References: Carl and Guiguet 1972, Chapman and Chapman 1975, Feldhamer et al. 1988, Moody et al. 1994, Nowak 1991, Wilson and Reeder 1993.

Moose

Alces alces

Other Common Names: American Moose, North American Elk.

Description

The Moose is the largest member of the deer family, and the males have the largest antlers of any living deer. The heaviest Moose on record is the Alaskan Moose of Alaska, weighing 595.5 kg for an adult male and 480 kg for a female. The Moose's body is about the size of a saddle horse and its legs are long and slender. Its hooves are large, elongated and sharp pointed, with well-developed lateral hooves or dew claws. The head is long with large ears, relatively small eyes and a square-lipped, bulbous muzzle (the upper lip overhangs). The nose is covered in short fur except for a small bare patch between the nostrils. A long (152–762 mm) flap of skin called a dewlap or bell hangs from the top of the throat. The shape of the bell viewed from the side can vary from slender to satchel (bulbous); while the satchel shape is typical of adult males, other shapes of the bell cannot be used to sex an individual. The neck is relatively short, and there is a noticeable hump above the shoulders. The tail is short and is not surrounded by a light rump patch, although most females have a small, usually indistinct area of white or lighter coloured hairs around the ano-genital region.

In general, the body, belly and rump are dark to blackish brown, sometimes greyish brown. The legs are lighter, often greyish white on the lower parts, especially on the insides and on the rear of hind legs. This is most evident in males. The face can be the same colour as the rest of the body, or lighter, or grey. The neck and top of the shoulders are covered in long hair (up to 150 mm) that is usually darker than the rest of the body, and can form a noticeable cape. Moose have antorbital, tarsal and interdigital glands and possibly a gland at the base of the bell. There is some disagreement whether they have metatarsal glands. Female Moose, like most deer, have four functional teats.

Moose calves up to about three months old are light red to reddish brown and lack spots. The hair around the eyes, on the muzzle and on the ears is black, while the inguinal area and the insides of the ears are light. At two to three months of age, the coat colour changes to resemble the adults.

The antlers of adult males are characteristically large, laterally oriented and palmated (palm-shaped). The antlers show their greatest

profile when seen from the front, and have one or two sections. When only one section is present, the stout main beam leaves the pedicel laterally and horizontally, before turning back and becoming heavily palmated. Tines of varying lengths grow along the outer edge of the palmated area. When the antler consists of two parts, the short main beam again leads from the pedicel, and then branches into an anterior section. This can consist of two or three large tines directed forward, or it can be palmated with outer tines along the edge. The posterior section is larger and always heavily palmated with a fringe of medium to short tines along the outer edge. Antlers of yearling males usually begin as a simple fork, but by the second year a third point or a small palm develops. Antlers are shed from late November to early January, with older males losing theirs first.

A Moose skull is long and narrow, especially the premaxillae and maxillae region, which accounts for over 36 per cent of its basilar length. The nasal bones on the other hand are short and the anterior nasal opening is large. There is a clear swelling of the frontal bones

just behind the orbits, and in front of this, a small depression along the centre line of the cranium (figure 48).

Measurements:
See separate subspecies descriptions.

Dental Formula:

incisors: 0/3
canines: 0/1
premolars: 3/3
molars: 3/3

Identification:
The Moose's unique features are: large size; dark brown coat; long, usually light-coloured, slender legs; large ears; long head with a bulbous muzzle; skin flap hanging from the throat; hump above the shoulders; and lack of an obvious light rump patch. In males, the large, broad palmated antlers bordered by short tines or points are also species specific. The European Fallow Deer, the only other species with palmated antlers, is much smaller, has smaller antlers and has different body coloration.

Elk and horses are similar in size to Moose, but both lack the shoulder hump and bell. Elk do not have the long head and bulbous muzzle, and have a large light brown or tan rump patch. The colour of the neck and legs in Elk is usually darker than the body, unlike Moose. A horse's legs are shorter and more stout than a Moose's, and have wide, single hooves. The long upper-neck mane and long-haired tail are unique to horses.

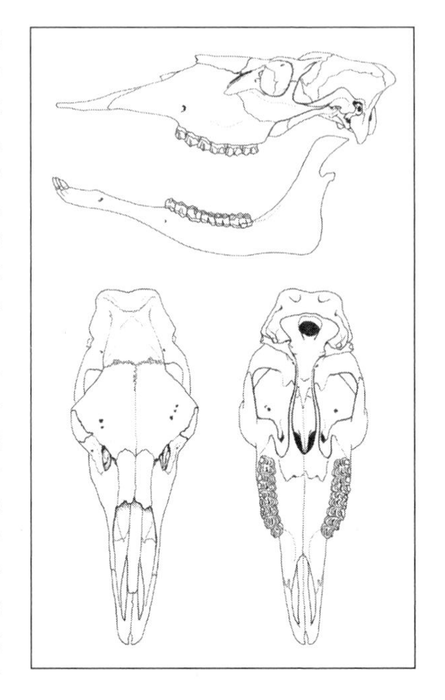

The skull of a Moose is unlike that of any other British Columbian ungulate. It has short nasal bones and characteristically long premaxillae and maxillae bones that extend more than 200 mm in front of the first upper premolar in adults. Similarly, the diastema of the lower jaw is long (only the Giraffe has a proportionately longer one).

Adult Moose tracks are unique, being much longer (about 150 to 180 mm) than those of Mule Deer or White-tailed Deer, and narrower with much sharper front points than similar-sized tracks of Elk, Caribou, Bison or Domestic Cattle (figure 38). Dry fecal pellets of Moose are much larger than those of Caribou, Mule Deer or White-tailed Deer. They are similar in size to those of Elk, but are rounded at both ends, whereas Elk pellets have a dimple at one end and a short projection at the other.

Natural History

The Moose is primarily a forest-dwelling ungulate and is especially common in boreal forests throughout their circumpolar range. Boreal forests are ecologically dynamic due primarily to forest fires. Unlike Caribou, Moose are well adapted to this changing forest landscape and are able to obtain the most nutritious food in northern and subalpine forests. For example, their diet consist largely of shrubs and young trees that are characteristic of young or regenerating forests. Under superior conditions, when fires create abundant and nutritious forage, Moose are able to produce twins and even triplets. Moose also make extensive use of river valleys where seasonal scouring by floods and ice create a complex pattern of young and old forest stands.

Avalanches on steep side slopes can result in similar vegetation diversity. Whether created by fire, logging, snow slides or seasonal flooding, successional stands with palatable browse species are important for Moose. Typically, winter is the season of greatest stress for Moose. At this time, they seek areas with reduced snowfall such as river valleys and areas of dense conifers where the canopy intercepts falling snow. There, access to food is comparatively better and shallow snow makes travel easier. In mountainous terrain, males often winter at the treeline where strong winds reduce the snow cover and Subalpine Fir provide an abundant supply of food.

Moose are primarily browsers, but their diet varies seasonally and geographically. In early winter, they feed on twigs of deciduous trees and shrubs. While it is difficult to generalize because of geographic differences, Moose commonly eat Paper Birch, willows and Trembling Aspen, with willows probably being the most important winter forage. Other preferred but less common foods include Red-osier Dogwood, Mountain-ash, Black Cottonwood, False Box, Saskatoon, Hazelnut and Highbush-cranberry. As winter progresses and snow continues to accumulate, covering many deciduous species, Moose feed increasingly on coniferous trees, particularly Subalpine Fir. Signs of Moose feeding on deciduous trees and shrubs are quite common. They range from heavily browsed shrubs that have short, dense twigs and branches, to young deciduous trees with their tops bent over or broken. Because of their long legs, Moose may sometimes kneel down to reach low-growing plants. At other times, they will straddle and push down small deciduous trees to gain access to the top branches. In spring and summer, they feed on new leaves and growing shoots of browse species and also take a wide range of herbs, such as Fireweed. They will also feed, sometimes heavily, on emergent and floating vegetation (e.g., Swamp Horsetail, bur-reeds and pondweeds) in swamps and ponds, and along lake margins. While consuming aquatic vegetation, Moose often venture into deep water up to their shoulders and feed with their heads completely submerged. Their use of aquatic plants in summer has been linked to these plants' high sodium content. In the late summer and fall, as forbs die back, Moose feed once more on deciduous browse species. An adult Moose is estimated to need as much as 19 kg of food every day.

Moose may be the least social of BC's ungulates. Like most hoofed mammals, the sexes live apart for much of the year, and the males are generally solitary. Females usually live only with their offspring until the young is just over a year old, although they may remain together longer if the mother fails to give birth that year. In some areas, males aggregate in the fall at the beginning of the mating season. Moose are particularly vocal during the rut, and a variety of vocalizations have been described as moans, barks, croaks and grunts. Females also vocalize in the rut, and also use a grunt to call the young. The young will sometimes bawl like a domestic calf.

Both sexes show aggression by lowering their ears with the insides facing forward and raising the hair on their neck and shoulders. In the rut, male Moose display their antlers when they approach each other, walking in a slow, stiff-legged gait, tilting their head and antlers from side to side in an exaggerated see-saw motion, sometimes accompanied by a nasal vocalization. The stiff gait and rhythmic tilting movement of the head may draw attention to the large, light-coloured palms that are oriented forward and toward the opponent. Two males will often circle each other, displaying in this fashion. Like many other deer, they will also thrash nearby bushes with their antlers during the interaction's early phase. Thrashing noises probably function as a threat; they may also attract rival males. One or both interacting males may show displacement feeding, turning away from the other to feed briefly or just lower the head to the ground. If one rival does not move away, the two opponents will likely lock antlers, pushing and wrestling until one finally breaks away. Fights occasionally become vicious and animals can be injured or even killed. Broken ribs are not uncommon. While there are substantiated reports of Moose locking antlers and dying because they are unable to disengage, this situation is extremely rare. Females in the rut are also aggressive, charging each other with ears down and hair raised, and striking out with a foreleg. While the antlers are still growing or after they have been shed, male Moose will fight using their front feet in the same manner as females.

Mating takes place between early September and late October in BC. At this time both sexes of this mostly solitary species will actively seek mates. The mating system of Moose is quite flexible, and ranges from pairs to loose associations of up to 20 adults. Habitat structure and Moose abundance seem to be the determining factors, with open areas allowing the larger associations. Female Moose appear much more proactive in the mating season than females of other ungulate species. They are also vocal at this time, making long quavering moans variously described as a long drawn out "oo-oo-aw" or "mwar" in which the last note drops. These calls can be heard more than three kilometres away.

A male Moose uses his forelegs to scrape a small depression in soft, wet ground. He then urinates and wallows in it, rubbing its bell, chin and lower parts of the antlers in the urine-soaked mud. The odour seems to attract females who will approach, sniff the male and then also wallow in the depressions. The male's courtship is not particularly elaborate. He approaches the female from her rear, sniffs her, and if she urinates, he almost always lip-curls. If she is coming into heat, he follows behind her and later places his chin on her rump. If she does not move away, he briefly mounts and copulates with her. Throughout his courtship, the male utters soft grunts or croaks, although the females reportedly show no obvious reaction to them. Females come into estrus every 21 days and remain in heat for about 24 hours.

Birth takes place in June after a gestation period of 240 to 246 days; newborns weigh from 11 to 16 kg. Twins are common in areas with good nutritional conditions, otherwise single births are the rule; triplets are rare. Females seek seclusion to give birth (e.g., islands in rivers). Mothers will drive away their yearling offspring, but may reunite with them a few weeks after giving birth. It is not clear whether young Moose use the follower or hider strategy. Predation on young Moose can be high in the first year of life, mainly by bears and Grey Wolves. Female Moose are extremely protective of their young and respond aggressively toward any perceived threat to their offspring, charging with ears back and hair raised, ready to lash out with their long front legs and sharp hooves.

If nutritional conditions are favourable, female Moose may give birth as early as their second birthday, but most are usually three years old before their first young are born. Yearling males produce viable spermatozoa but have little opportunity to mate until a few years older. Determining the age of a Moose is most easily accomplished by counting the cementum annuli of the incisors. The average life expectancy for adult Moose is 8 to 10 years, and few Moose in BC are older than 17 years of age.

Wolves are the main predators of Moose, most of the time taking the oldest and youngest members of the population, while bears prey heavily on young in some areas. Most healthy adult Moose can defend themselves from predators with aggressive behaviour, including kicks from their front and hind feet. In some regions of the province, collisions with trains can be a major cause of winter mortality. This occurs when Moose become trapped in the railroad right-of-way by high snow banks piled alongside the tracks.

Some Moose can suffer severe hair loss in the late winter or spring caused by heavy infestations of the Winter Tick. Affected individuals rub and scratch themselves in an effort to rid themselves of the parasites and in the process, remove their outer guard hairs, leaving the lighter coloured under-hair. The resulting light colouring has led to the term Ghost Moose. The Winter Tick is carried by other species besides Moose, including Caribou, Elk, Mule Deer, White-tailed Deer and Bison. The ticks stay on the host during the winter, then in the spring, the fertilized females drop off the host to lay their eggs. Larval ticks seek a host in the fall, gradually developing into adults during the winter. If tick numbers and hair loss are extreme, affected Moose may become anaemic from blood loss, and in extreme cases, die from exposure and hypothermia in times of inclement weather. Moose in BC are also infected by the American Liver Fluke. If infestations are heavy and associated with malnutrition, they may contribute to the death of the animal. Moose Sickness is a neurological disease caused by the Meningeal Worm. The usual vertebrate host is the White-tailed Deer. Transmission of this parasite from deer to Moose has been implicated in population declines in eastern North America in the first half of the 20th century, but the true role of the parasite is unknown. It is not reported in western Moose populations.

Northwestern Moose
Alces alces andersoni

Other Common Names: Anderson's Moose, British Columbia Moose, Western Canadian Moose.

Description

Externally, Northwestern Moose differs little from the other two subspecies in the province. The cranial dimensions show extensive overlap, and it is possible that Northwestern Moose are somewhat intermediate in colour between Alaskan Moose and Yellowstone Moose. See the Description for the species, above.

Measurements:
weight -
  male: 437.3 kg (339-519) n=4
  female: 312.3 kg (270-352) n=3
total length -
  male: 3,085 mm (2,712-3,454) n=15
  female: 2,899 mm (2,410-3,251) n=85

tail vertebrae -
male: 133 mm (95-171) n=2
female: 87 mm (30-80) n=26
hind foot length -
male: 804 mm (775-832) n=2
female: 856 mm (800-890) n=28
shoulder height -
male: 1,877 mm (1,924-1,829) n=2
female: 1,651 mm, n=1
ear length -
male: 251 mm (248-251) n=2
chest girth -
male: 2,145.3 mm (1,981-2,489) n=13
female: 2,012 mm (1,118-2,489) n=84
skull length -
male: 530.5 (516-551) n=4
female: 545.0 mm (538-554) n=3
skull width -
male: 210.0 mm (198-220) n=4
female: 199.3 mm (195-203) n=3

Range

Northwestern Moose has the widest distribution of the three subspecies in BC. It is found throughout most of the province, except the extreme northwest and southeast corners. Northwestern Moose is also generally absent west of the Coast Mountains, although it is found in some small areas such as along the Skeena River and at the heads of major inlets such as Butte, Douglas, Knight and Portland. It does not occur on either Vancouver Island or Haida Gwaii.

Good places to see Northwestern Moose include many areas around Prince George, the Kootenay River valley in Kootenay National Park, Wells Gray Provincial Park and the Elaho River valley northwest of Squamish.

Conservation Status

Between 140,000 and 235,000 Moose lived in British Columbia in 2011. Most were Northwestern Moose, which is not considered to be threatened. Some of the highest densities are reported around Prince George, near Horsefly, along the Interior Plateau and on the east side of the Rocky Mountains northwest of Fort St John.

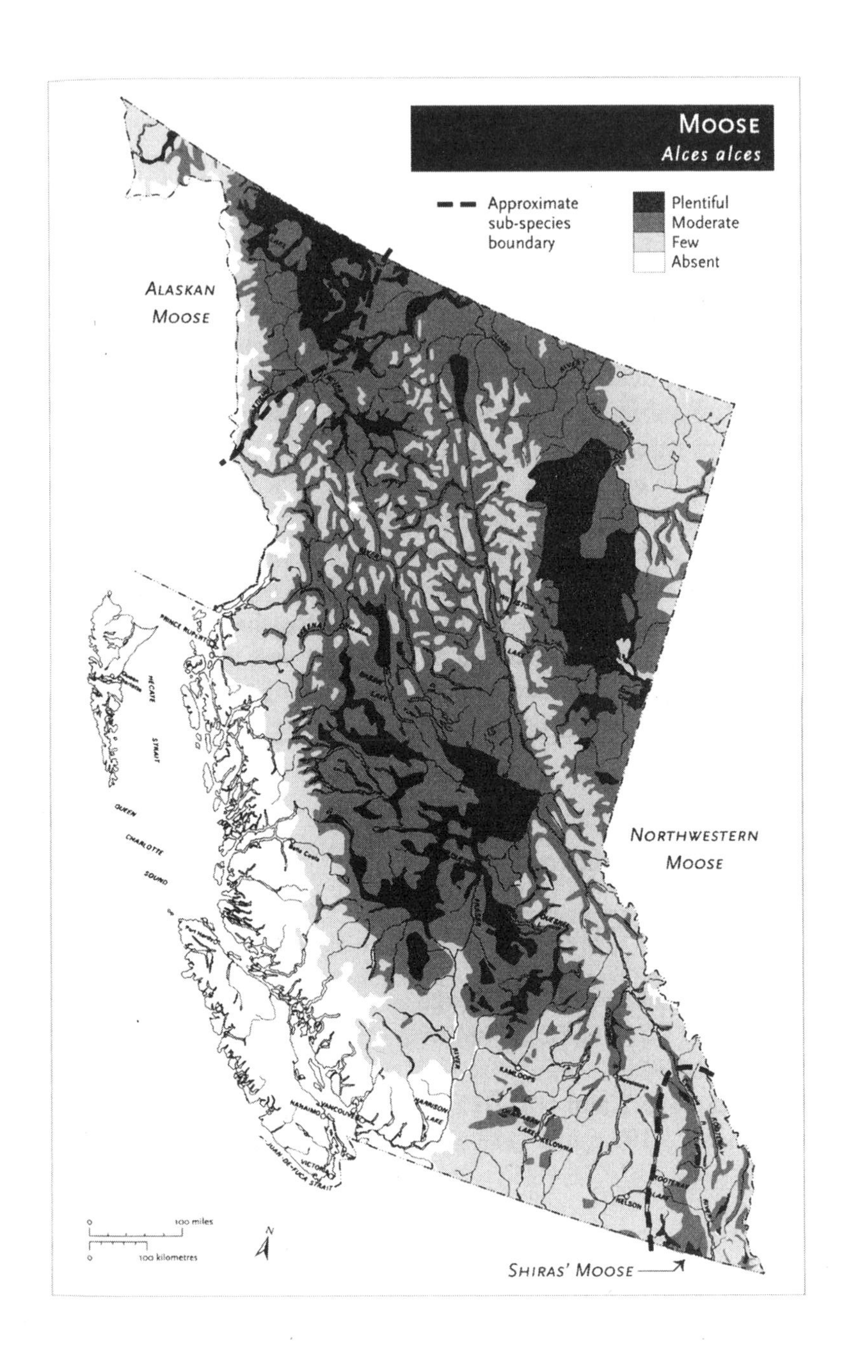
Moose
Alces alces
Approximate sub-species boundary
Plentiful
Moderate
Few
Absent
Alaskan Moose
Northwestern Moose
Shiras' Moose
Prince Rupert
Hecate Strait
Queen Charlotte Sound
Kamloops
Kelowna
Nelson
Vancouver
Victoria
Nanaimo
Juan de Fuca Strait
Harrison Lake
Kootenay Lake
100 miles
100 kilometres

Alaskan Moose

Alces alces gigas

Other Common Names: None.

Description

Alaskan Moose is the largest subspecies of Moose, although whether it is larger than Northwestern Moose in nearby northern populations is uncertain. In terms of body colour, it is supposed to be the darkest of the province's three subspecies.

Measurements:

weight -
- male: 505 kg (492-517) n=2
- female: 375 kg (263-451) n=11

total length -
- male: 3,055 mm (?) n=5
- female: 2,495.5 (2,413-2,578) n=2

tail vertebrae -
- male: 178 mm (?) n=2
- female: 200 (197-203) n=2

hind foot length -
- male: 838 mm (?) n=2
- female: 794 (787-800) n=2

shoulder height -
- male: 1,867 mm (?) n=2
- female: 1,820 (1,798-1,842) n=2

chest girth -
- male: 2041 mm (?) n=5
- female: 2013 mm (?) n=23

skull length -
- male: 557.0 mm (544-570) n=2
- female: 523.0 mm (517-530) n=4

skull width -
- male: 214.3 mm (208-219) n=3
- female: 193.3 mm (188-204) n=4

Range

In *The Mammals of British Columbia* (1965) Cowan and Guiguet state that Alaskan Moose lives only in the extreme northwest corner of the province "from Telegraph Creek north through the Yukon drainage basin including Atlin and Teslin lakes". The southern and eastern boundaries of its range are uncertain, so the population of this subspecies is difficult to estimate. Some of the highest densities are along the Tatshenshini and Teslin rivers, and around Teslin and Atlin lakes.

Conservation Status

There is no separate estimate of the number of this subspecies, but it is not considered threatened in the province.

Shiras' Moose

Alces alces shirasi

Other Common Names: Yellowstone Moose, Wyoming Moose.

Description

Data on Shiras' Moose are insufficient to give an accurate description. Shiras' Moose is supposed to be the smallest subspecies in British Columbia. Certainly, its antlers are smaller than those of the other two subspecies; and they are usually less palmated, but this pole-horn type, as it is sometimes referred to, is found in other subspecies, so is not unique to Shiras' Moose. The coat colour of this moose is generally lighter than the other subspecies; it has been described as pale rusty yellowish brown or pale brown along the back and upper sides of the neck, shading to brown-black on the lower part of the body, except on the lower belly, which is a pale buff colour. The ears are described as light grey and the hooves small. Cranial dimensions show much overlap among the subspecies, so measurements are not useful criteria for separating them.

Measurements:

weight -
- male: 405 kg (311-524) n=20
- female: 338 kg (272-386) n=7

total length -
- male: 2,540 mm n=1

hind foot length -
- male: 762 mm n=1

skull width -
male: 195.7 mm (181-212) n=10
female: 193.6 mm (188-197) n=5

Range

Shiras' Moose is restricted to an area in the extreme southeastern part of the province, although the exact northern border is debatable. The maximum distribution is considered to be south of the Trans-Canada Highway ranging from the upper Flathead River valley, east to the adjacent west slopes of the Rocky Mountains, and south to the international border. The western boundary is also uncertain, but may be the east side of Kootenay Lake. In *The Mammals of British Columbia* (1965), Cowan and Guiguet describe a much more restricted distribution area for this subspecies: "the Flathead Valley and adjacent Rocky Mountain areas south of Crowsnest Pass". The Flathead and Elk river valleys provide the best places to see this subspecies.

Conservation Status

In 2011, there were between 7,000 and 9,000 Moose in the Kootenay region, many of which are probably Shiras' Moose. This subspecies is not considered at risk.

Taxonomy

In his 1952 monograph, Randolph Peterson used mainly size differences of the skull to separate the various subspecies of Moose, even though there was much overlap. Today such differences are not considered particularly useful, and Moose taxonomy needs to be reexamined. The presence of Shiras' Moose in British Columbia represents the northern limit of this subspecies' range in North America. Peterson thought that most, if not all, of Shiras' Moose in the province were intergraded with Northwestern Moose and that individuals in southeastern BC had intermediate characteristics between the two subspecies. Shiras' Moose is named after George Shiras III, who reported this Moose in Yellowstone National Park between 1908 and 1910. The type locality for Alaskan Moose is Tustomena Lake, Kenai Peninsula, Alaska; for Northwestern Moose it is Sprucewood Forest Reserve, 24 km east of Brandon Manitoba; and for Shiras' Moose it is the Snake River, Wyoming.

In 1998, a well-preserved Moose antler was found in a peat bog near Smithers. Its radiocarbon date of over 5,510 years before present makes it one of the earliest known fossil Moose in BC.

Traditional Aboriginal Use

Moose were – and still are – an important species for both meat and hides to many First Peoples in BC, especially for north coastal and northern groups before the 1900s. Aboriginal people used Moose bones to make tools: the shoulder blades for scrapers, the lower jaws as clubs, and in the northern interior, the fibula as an awl. They made a variety of articles from the hides, including moccasins, clothing, pouches, bags and tumplines (for carrying heavy loads). Moose skins were important trade items – people on the north coast obtained them from the interior. Some First Nations used untanned hides to make body armour.

Remarks

Alces is Latin for "elk", *Alce* is Greek for "elk", and in German it is *elch*. The common name Moose is derived from the Algonkian word *musee* meaning "eater of twigs".

Moose numbers appear to have fluctuated in the province, at least since the arrival of the first Europeans. But there is some uncertainty about the historical distribution of this species in BC. Research by Dr James Hatter suggested that historically, Moose were absent from much of central BC and did not occupy this region until the 1900s. At this time, European settlers and miners cleared dense forests, thereby creating Moose habitat and allowing their southerly expansion. In 1931, Dr Ian MacTaggart Cowan was told by members of the Tsilhqot'in (Chilcotin) First Nation that Moose first moved into their area about five or six years earlier, and that they had no word for Moose in their language. Dr David Spalding, in his 1990 report, found evidence suggesting that some Moose may have always been present in much of BC, except for west of the Coast Range, even if only at low densities. It is reasonably clear, however, that Moose numbers did increase and that they expanded southward after the turn of the century. It also appears that this southerly expansion was followed shortly after by an increase in numbers of Grey Wolves. Together, these two factors have been blamed for the demise of Woodland Caribou in the 1930s.

Moose, particularly in the northern parts of their range, are well adapted to the normal dynamics of the boreal forest, but logging is increasingly replacing fire as the primary agent of forest change, and thus the habitat requirements of Moose need to be considered in forest harvesting plans.

At certain times of the year, Moose can be one of the most dangerous large mammals in North America. Females with young in early spring should not be approached, and both sexes are best avoided during the mating season. Usually, the first signs of aggression are lowering the ears, followed by raising the hair on the neck and shoulders. Both males and females use their long front legs and sharp hooves to strike out at other Moose, as well as at predators and humans.

Selected References: Blood et al. 1967, Franzmann and Schwartz 1998, Geist 1963, Hatter 1950, Kunkel and Pletcher 2000, Lankester and Samuel 1998, Lent 1974, Peek 1974, Peterson 1952, Samuel 1993, Spalding 1990.

Mule Deer

Odocoileus hemionus

Other Common Names: None.

Description

Mule Deer is a medium-sized deer with a stocky body, slender legs, and a medium-length tail. Its common name stems from its characteristically large ears, which are light coloured on the inside with a dark brown-black rim. The coat is greyish brown in winter, and reddish brown in summer. The belly is the same colour as the rest of the body, but the insides of the legs are lighter. There is a dark brown cap on the forehead extending from the ears to between the eyes; in older males, the cap is often darker along the sides just above the eyes and along the front edge. In adult males, this cap is especially noticeable because the face below it can be light grey to almost white. The cap usually does not reach the top of the eye, so there may be a lighter

eye-ring above the eye. In summer coat, the cap is not always obvious. The nose is black and mainly hairless, sometimes with a narrow light or white area at the posterior edge, and the chin is also white with a vertical black stripe toward the back of the lips. There is always one white patch at the top of the neck below the throat, and sometimes a second one below it. The rump patch is white but varies in size with the subspecies, as does the colour of the tail (figure 41). Mule Deer have poorly developed frontal skin glands located in the forehead. The metatarsal glands on the outside of the hind legs are large and well developed, as are the tarsal glands on the inner sides. Both pairs of these hind-leg glands are recognizable by the tufts of longer hair, sometimes of a darker colour, that mark their position. Other paired glands include antorbitals and interdigitals on each foot.

Young of the year are a reddish brown with white spots usually in horizontal rows along either side of the unbroken white mid-dorsal stripe. They also have a small light rump patch, and the hair inside the ears is light. At about 2.5 to 3 months of age, the coat becomes similar in colour to that of the adults.

The antlers of adult male Mule Deer show a bifurcated pattern: after the small brow tine, there are two pairs of relatively even forks, typically making a total of five tines on each antler. Atypical antlers are occasionally found on Rocky Mountain Mule Deer in BC, and as many as 48 tines have been counted on the antlers of one animal. Atypical antlers are rarely found in either of the two Black-tailed subspecies. The antorbital depression (lachrymal pit) on the skull is relatively deep.

Measurements:
See separate subspecies descriptions.

Dental Formula:

incisors: 0/3
canines: 0/1
premolars: 3/3
molars: 3/3

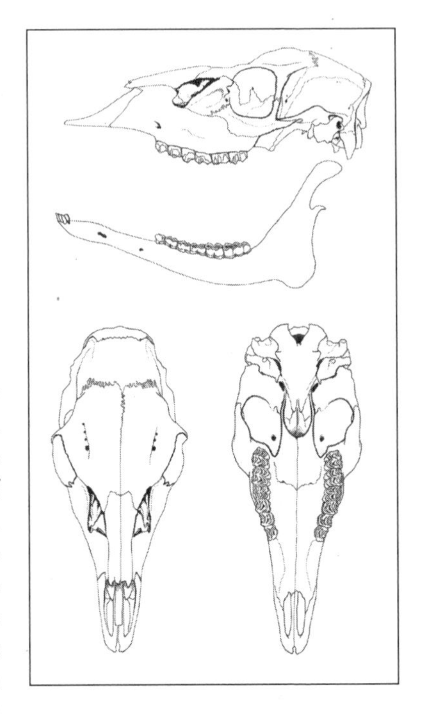

Identification:
The Mule Deer is similar in size to the White-tailed Deer and European Fallow Deer. Unlike White-tailed Deer, Mule Deer of both sexes have a dark forehead patch, their metatarsal glands lack the white hairs and the rump patches differ markedly (figure 41). The Rocky Mountain Mule Deer's large white rump patch and short-haired white tail with its black tip make it easily distinguishable from the White-tailed Deer, which has a long tail with long hairs that are reddish-brown on the upper surface and white on the lower. The upper side of the tail of both subspecies of Black-tailed Deer is dark brown or black. In males of all three subspecies of Mule Deer, the forked tines that grow from the main beam contrast with the unforked tines of White-tailed Deer (figure 42). The skull is similar in size to those of White-tailed Deer and Fallow Deer, but the antorbital depression of Mule Deer is much deeper than that of White-tailed Deer, and unlike the Fallow Deer, the posterior nares are divided by the vomer (figure 49).

Adult Mule Deer lack the spots of the spotted form of Fallow Deer, and compared to the two other colour phases, Mule Deer are either darker than the light phase, or lighter than the dark phase. Also, compared to all colour phases of Fallow Deer, the Mule Deer's dark forehead extends forward past the eyes, and the ears are proportionately larger and have dark rims. The Mule Deer's rump patch lacks the black side stripes that the Fallow Deer's has. The antlers of male Mule Deer are not palmated as is typical of adult Fallow Deer.

Tracks are not easily distinguishable from other medium-sized deer; but with practice, they can be separated from those of Mountain

Goat and Bighorn Sheep, because deer tracks are more pointed and delicate in shape (figure 38). Feces can be difficult to distinguish from those of similar-sized ungulates.

Natural History

Mule Deer is a widespread and adaptable species that inhabits a broad range of habitats. But the two coastal subspecies prefer different habitats than the Rocky Mountain Mule Deer of the interior.

In general, the Rocky Mountain Mule Deer prefers open forested areas or parklands with adjacent grasslands, but also inhabits timbered (Douglas-fir, Ponderosa Pine) river breaks, foothills and mountain slopes. In the interior, Rocky Mountain Mule Deer is often associated with the Interior Douglas-fir Zone, especially in winter. The open grasslands and underbrush in adjacent mature timber (80 to 140 years old) provide food, while the forest supplies cover from both predators and weather. In winter, this deer also prefers southerly or westerly exposures below 1,500 metres elevation with gentle slopes. In the central interior of British Columbia, it prefers old forest stands in winter, especially during periods with moderate (250 to 400 mm) and deep (more than 400 mm) snow. A moderate to dense forest canopy is characteristic of deer winter range during periods with deep snow.

In contrast, both Columbian Black-tailed Deer and Sitka Black-tailed Deer inhabit the dense coastal rainforests. But they also need access to open habitats such as riparian, subalpine and alpine areas, early successional stages of forest, and the underbrush of old-growth forest for the forages that grow in them. The coastal forests also lessen the effects of deep snow. A dense forest canopy not only provides a litter of lichens and twigs for winter food, but also intercepts much of the snowfall. Because less snow accumulates on the forest floor, shrubs are more accessible and travel is easier for the deer. Deer try to avoid areas where they begin to sink almost half way to their chest, because the energy costs of moving increase dramatically at this point. It can also be much harder to escape predators in deep snow. Where there are no alternatives, deer may yard, meaning they remain in the same general area and use snow trails that develop into open runways through repeated use.

A dense forest canopy also affords important thermal cover for deer, providing cooling shade in summer and shelter in winter. The two coastal subspecies seem to prefer forest edges where they have ready access to both open and closed habitats. This is similar to habitat preferred by Rocky Mountain Mule Deer, although the interior deer often move further from cover. All three subspecies like to bed down in forest and shrub edges along natural or man-made openings and grasslands, or along ridge tops and mountain slopes. These areas not only provide concealment but also a clear view for predators.

Although individuals vary, most deer make seasonal altitudinal migrations, moving to higher elevations in summer. Rocky Mountain Mule Deer may travel up to 60 km between summer and winter ranges, with some moving as much as 120 km. In spring, these deer move from their winter range to wetter areas with a greater variety and biomass of forages available in summer, thus maximizing their nutritional opportunities. In fall and early winter, snow accumulation at the higher elevations forces the deer back to the lower valleys. Columbian Black-tailed Deer seem to move less than Mule Deer, restricting their seasonal migrations to within watersheds. All three subspecies are excellent swimmers and both the Columbian and Sitka subspecies are known to swim between islands along the coast.

Although Mule Deer are primarily browsing herbivores, they eat a variety of plant foods, apparently related to local conditions. Rocky Mountain Mule Deer seem to eat more grasses than their coastal relatives. All three species in British Columbia prefer browsing on Douglas-fir, Saskatoon and willows; they also eat many species of forbs, such as Fireweed, and a variety of grasses. In some areas in the late fall and early winter, Rocky Mountain Mule Deer will paw through the snow to eat the brown, dead remains of large leafy forbs, such as Cow-parsnip, that have turned into a natural silage. Important plant foods for Columbian Black-tailed Deer in spring and summer include forbs such as Fireweed and Pearly Everlasting, ferns such as Bracken, and browse species like blackberry, Douglas-fir, raspberry, Salal, Salmonberry, Thimbleberry and willows. In winter, important forages are Douglas-fir, Red Huckleberry, Salal and Western Red-cedar, as well as Deer Fern and arboreal lichens (e.g., *Alectoria*, *Bryoria* and *Usnea*). Research has shown that the normally low digestibility of Salal is improved when eaten with other plant species. Little is known about the ecology or behaviour of Sitka Black-tailed Deer in BC.

Adult males and females of all three subspecies of Mule Deer live apart for most of the year, often in groups. During winter, Rocky Mountain Mule Deer may form larger groups than the other two subspecies, especially late in the season and in open habitats, where groups of 60 or more can assemble. In Columbian Black-tailed Deer, females are less social than males. Their basic social unit, as in the Rocky Mountain subspecies, is the family group, consisting of a mother with her offspring of the current year and often those of the previous year; two or more maternal units sometimes join to former larger groupings. Even larger aggregations can form when deer feed in clear-cuts, but these are usually only temporary and show little if any cohesion. During spring in some areas, the more social Black-tailed males are often seen in groups of 20 or more, but divide into smaller, looser-knit units in summers. During periods of high predation, the large spring groups of Columbian Black-tailed Deer are seldom seen.

When alarmed, Mule Deer will stamp a front foot, snort and, sometimes, walk a few paces in a stiff-legged gait. In the open, they often flee using a distinctive gait called stotting, where the animal bounds along as if it has springs on its feet, moving surprisingly fast. Stotting is more common in the Rocky Mountain Mule Deer than in Columbian Black-tailed Deer, and less so in Sitka Black-tailed Deer. Unlike White-tailed Deer, Rocky Mountain Mule Deer do not raise their tail upright and wave it when alarmed, although Black-tailed Deer, especially a female with young, will sometimes do so. Where thick cover is available, Black-tailed Deer prefer to sneak away from predators.

Unlike most other ungulates, male and female Mule Deer urinate alike, lowering their hind legs and crouching. Both sexes also perform hock-rubbing (or hock-urination), in which they stand with hind legs together, rubbing their tarsal glands and urinating on them (figure 18). Hock-rubbing is done year-round, but during the rut it is done primarily by large males, often when encountering another male or after a bout of antler thrashing. Rubbing their antlers on shrubs, bushes and small trees is a common behaviour of males to get rid of the velvet, and also in the rut when it may serve as both visual and auditory signals. It may also leave an olfactory message if the forehead glands are rubbed against the vegetation.

When meeting in the rut, similar-sized large males approach each other with heads low, ears back, tails held horizontally or moving rapidly up and down, and body hair erect. They will also bush-thrash and hock-rub as they approach. When closer, the males circle without looking directly at each other, sometimes in a crouch. Males also make a rut-snort (a pig-like noise) while snapping the head up and then immediately lowering it again and hissing for 5 to 10 seconds. Rut-snorts of Columbian Black-tailed Deer are reportedly louder than those of Rocky Mountain Mule Deer and are followed by a series of loud grunts. At this point in an interaction, one male may move away, and if he does, his opponent will often follow him for some distance.

Sparring matches are not uncommon between males of different sizes, but only similar-sized males will engage in a real fight, presumably to determine dominance. Even so, these real fights are rare and are preceded by the displays described earlier. Fights can be violent – it is not unusual for one of the combatants to be injured. The animals engage antlers, then twist and shove vigorously until one disengages. The one that breaks off flees as quickly as possible to avoid being attacked by the winner, who chases him, striking the ground with his forefeet. Rocky Mountain Mule Deer males also make a barking call during these chases. Locking antlers is extremely rare – but if it occurs, both combatants, unable to disengage, may starve to death. Females, and males without hard antlers, fight by striking out with their front legs, either from a normal standing position or by rearing up on their hind legs. Often prior to striking out, the animal may show a head-high threat, raising the neck and head with the ears back.

The female Mule Deer's estrous cycle lasts for 24 to 36 hours, and the interval between heats is 20 to 29 days. The mating system is a tending pair, with one male guarding and courting a single female. When the male has copulated several times, he moves in search of another female. A male checks a female's reproductive status by sniffing her urine and lip-curling. To encourage her to urinate, the courting male moves toward her rear without looking directly at her, his head low and neck stretched, and makes a soft buzzing sound while flicking his tongue in and out, before finally nosing her rear. The female may move away a few steps or crouch and urinate for the male. A second quite different approach is when the male, again behind a female, makes a low drawn out bleat and then suddenly rushes toward her with his head lowered. He then strikes the ground with his front feet, at the same time making a loud coughing roar, before chasing after her. When the male stops, so does the female, and then she urinates. A female approaching estrus will feed with her tail raised slightly, and when she allows the male to lick her rear and put his chin and neck on her rump, the male will try to copulate with her. Successful ejaculation occurs in a brief copulatory jump, with the male's hind legs off the ground and head thrown back. After this, the female bounds away a short distance, stands with her back arched and tail erect, and then contracts her belly several times. These contractions can go on for up to an hour after copulation. Columbian Black-tailed Deer begin mating in late October on the southern part of Vancouver Island and on the Gulf Islands, and later further north, extending into November and early December. Rocky Mountain Mule Deer usually start in mid November and continue until mid December.

The average gestation period is 203 days. The female bears one or two young, each weighing 2 to 4 kg, usually in June. Twins are quite common, except on poor range, and triplets are rare. The young are hiders and remain hidden while their mothers feed elsewhere, sometimes in small temporary groups. The spotted brown coat of the young provides camouflage as it lies in the undergrowth.

Mule Deer reach sexual maturity at about 18 months of age, and most females give birth for the first time on their second birthday, usually to a single young. Males probably have their first chance to participate in the rut at three or four years of age. The pelvis can be used to determine the sex of a skeleton where the head is missing. In adult male Mule Deer, there is a protuberance called the suspensory tuberosity where the penis ligament attaches to the anterior edge of each ilium just above their junction.

With experience, tracks of adult male and female Black-tailed Deer can be distinguished by their shape. Male tracks are longer and wider than those of the female. There are various methods for determining the age of Mule Deer. The most reliable are the tooth eruption sequence for animals up to about three years of age, and cementum annuli counts using incisors for older individuals. Probably few Mule Deer live longer than ten years and most live for no more than four or five.

Diseases and parasites do not seem to be major causes of death for Mule Deer. Like White-tailed Deer, Mule Deer are susceptible to epizootic haemorrhagic disease (EHD), but there have been no reports of this disease in BC's populations. Predation is probably the main mortality factor for Mule Deer, and severe winters can have devastating effects. Grey Wolves and Cougars are the major predators of all three subspecies. On the mainland, Coyotes are also important predators on young and probably also take some adults. Other less frequent predators are Bobcats, Lynx, and Wolverines, and both Black and Grizzly bears will take young deer in spring. Wolves from the mainland moved over to Vancouver Island in the early 1970s and increased in numbers. Besides swamping the indigenous Grey Wolf on Vancouver Island, they also drastically reduced Columbian Black-tailed Deer populations, especially on the northern part of the island. Deer numbers have only recently started to increase again.

Columbian Black-tailed Deer

Odocoileus hemionus columbianus

Other Common Names: Blacktail, Coastal Black-tailed Deer, Coast Deer.

Description

Columbian Black-tailed Deer is generally similar in coloration to the Rocky Mountain Mule Deer, but it is smaller and has a relatively short face and ears. The Columbian Black-tailed Deer's rump patch is also much smaller. The upper tail surface is black for about the last three-quarters of its length, with the basal quarter brown (figure 41). Its metatarsal gland, located about halfway along the metatarsal, is 64 to 70 mm long – about half as long as that of the Rocky Mountain Mule Deer, which is located closer to the proximal end of the metatarsal.

For most of us, it is probably impossible to tell Columbian Black-tailed Deer apart from Sitka Black-tailed Deer in the field where their distributions meet. The Columbian is less reddish brown in winter than the Sitka, but more red in summer, and it has a larger forehead patch. It also lacks the Sitka's obvious dark line down the nose. The Columbian's tail has more black on its upper surface than the Sitka's.

The antlers of adult male Columbian Black-tailed Deer are darker, smaller and less massive than those of Rocky Mountain Mule Deer, and they usually have fewer forks. Where these two subspecies overlap on the east side of the Coast Mountains, hybrids occur, and again the most useful distinguishing feature is probably the tail. The hybrid's tail usually has a black stripe running from base to tip; its other characteristics are intermediate. There may be small differences in antler shape between Columbian and Sitka Black-tailed Deer, but these are difficult to see in museum specimens, and even harder to see in the field. The maximum breadth of the sub-lachrymal extension of the jugal bone in the skull is more than 5 mm in Columbian Black-tailed Deer, compared to less than 5 mm in Sitka Black-tailed Deer.

Measurements:

weight -
- male: 54.5 kg (45-64) n=5
- female: 45.6 kg (?) n=17

total length -
- male: 1,614 mm (1,472-1,780) n=3
- female: 1,370 mm (?) n=14

tail vertebrae -
    male: 186 mm (162-205) n=3
    female: 140 mm (152-191)
hind foot length -
    male: 454 mm (410-495) n=4
    female: 413 mm (?) n=16
ear length -
    male: 142 mm n=1
    female: 155 mm n=1
chest girth -
    female: 805 mm (?) n=14
skull length -
    male: 230.7 mm (227-233) n=5
    female: 225.4 mm (216-230) n=5
skull width -
    male: 102.0 mm (102) n=3
    female: 97.2 mm (93-100) n=5

Range

Columbian Black-tailed Deer are found throughout Vancouver Island, on almost all the smaller islands, and along the coast on the west slopes of the Coast Mountains from the international border north to about Rivers Inlet. There it begins to intergrade with Sitka Black-tailed Deer. The eastern boundary of its distribution is approximately the crest of the Coast Mountains, and here too it intergrades with the other subspecies, Rocky Mountain Mule Deer.

On Vancouver Island, Columbian Black-tailed Deer can be found in almost any forested area where recent logging has taken place with adjacent unlogged habitat. The Sechelt Peninsula and most of the islands in the Strait of Georgia from Malcolm to Saturna islands are also areas of high density and thus good places to spot these deer. If undisturbed, Columbian Black-tailed Deer adapt well to the presence of humans; if there are forested areas, they can be found within many city parks, gardens and other green areas.

The effects of forestry on Columbian Black-tailed Deer have been well studied. Results show that as long as a well-interspersed mosaic of different forest types is maintained, along with adequate winter range for severe winters (i.e., old-growth forests), deer will thrive. In some cases, they have reached much higher densities than in the past. The successional stages following logging provide abundant forage, and with older timber nearby, there is the necessary thermal and security cover and relief from deep snow. On Vancouver Island, and probably

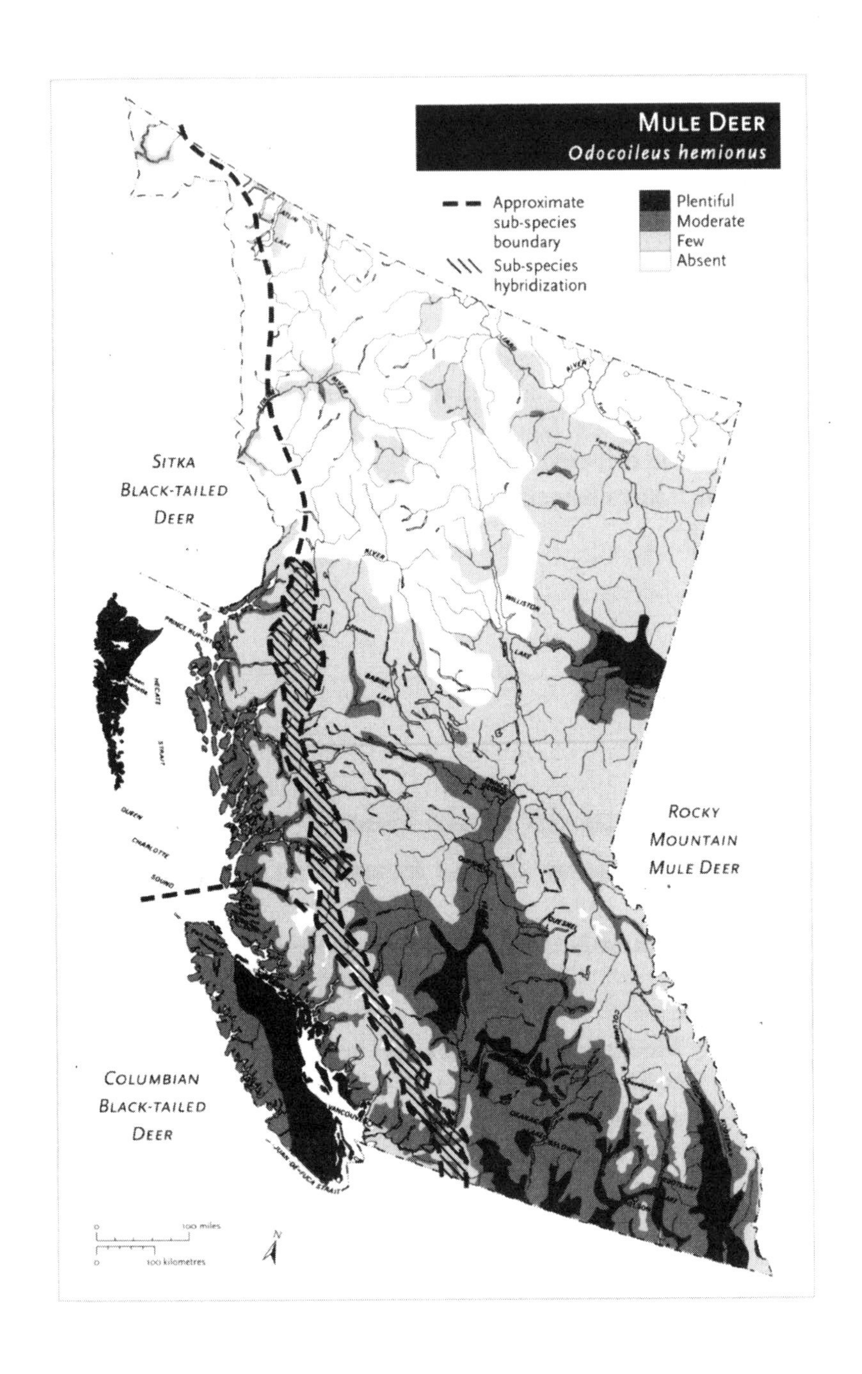
MULE DEER
Odocoileus hemionus
Approximate sub-species boundary
Sub-species hybridization
Plentiful
Moderate
Few
Absent
SITKA BLACK-TAILED DEER
ROCKY MOUNTAIN MULE DEER
COLUMBIAN BLACK-TAILED DEER

along all BC's coast, preserving intact old-growth forests at low to mid elevations is recommended as vital for maintaining or rebuilding Black-tailed Deer populations of both subspecies. This is important because individual Black-tailed Deer, like some other ungulates, show strong attachment to an area and to a habitat-use pattern, remaining there for most of its life. As a result, Black-tailed Deer are susceptible to sudden significant changes in habitat such as those caused by forest removal. Individuals are unable to respond quickly with new, more appropriate strategies when change occurs suddenly.

Conservation Status

Based on the 2011 estimates, there are 99,000 to 155,000 Black-tailed Deer (both subspecies) in BC. Prior to 1975, some of the highest densities on the island were on the western slopes around Gold River, Quatsino Sound and Tahsish Inlet, in the Nitinat-Alberni area, and on Nootka Island. Currently, the highest densities are on the east side of the island, often near human habitation. Columbian Black-tailed Deer are not threatened.

Rocky Mountain Mule Deer
Odocoileus hemionus hemionus

Other Common Names: Mule Deer, Interior Mule Deer, Muley.

Description

The Rocky Mountain Mule Deer's ears are large and its tail has relatively short white hairs, except on the tip where they are longer and black. Surrounding the tail is a large white rump patch, while the top of the rump just at the base of the tail is often darker than the rest of the body (figure 41). Rocky Mountain Mule Deer also tend to be more greyish brown in winter pelage than either of the two Black-tailed subspecies, and in some individuals the hind legs can be a more reddish brown than the rest of the body. As winter proceeds, the coat colour often becomes bleached to a lighter grey-brown. The metatarsal gland is over 100 mm long and is located close to the proximal end of the metatarsal. This is longer than the gland in either of the other two subspecies and its location on the metatarsal also differs. For other differences between the three subspecies of Mule Deer, see the Description for Columbian Black-tailed Deer.

Measurements:

weight -
  male: 103.7 kg (75-126) n=3
  female: 64.6 kg (48-80) n=10

total length -
  male: 1,778 mm (1,727-1,880) n=3
  female: 1,448 mm (1,270-1,549) n=3

tail vertebrae -
  male: 152 mm (135-178) n=3
  female: 178 mm (165-197) n=3

hind foot length -
  male: 584 mm (533-584) n=3
  female: 483 mm (457-508) n=3

skull length -
  male: 274.6 mm (273-284) n=5
  female: 250.5 mm (230-257) n=4

skull width -
  male: 120.0 mm (106-131) n=6
  female: 112.0 mm (98-121) n=4

Range

The extensive distribution of Rocky Mountain Mule Deer in BC stretches almost continuously from Manning Provincial Park to the Rocky Mountains. It extends northward along the east slopes of the Coast Mountains to about 56° North, bounded on the west by the crest of the mountains, then runs eastward to just south of Tchentlo and Chuchi lakes in north-central BC, before swinging north again to the south end of Williston Lake in the Rocky Mountain Trench. From there the distribution continues north along the east side of the Rocky Mountains as far as the Liard River, and east through the Peace River area to the Alberta border. There are other separate distribution areas with a few sparse populations of Rocky Mountain Mule Deer in the northern part of the province. These sparse populations are found at the north end of Williston Lake around the Ingenika and Finlay rivers south of Ware; along the Gataga, Kechika and Dall rivers at the north end of the Rocky Mountain Trench; north of Boya Lake Park; and on the Stikine Plateau along the upper Spatsizi and Ross rivers, between

Tuaton and Laslui lakes in the south and Cold Fish Lake in the north, extending upstream along the Stikine River from its junction with the Chutine River to around Snow Peak and south along the Klappan River to the beginning of the Little Klappan River. The most northerly distribution areas are east of the Coast Mountains around Atlin and Tusthi lakes, and on the north side of Graham Inlet.

Mule Deer can be seen almost anywhere within their range where densities are moderate to high, such as in central and south-central British Columbia.

Conservation Status

The 2011 estimate for Rocky Mountain Mule Deer in the province was 115,000 to 205,000 animals, with almost half in the Thompson-Nicola and Okanagan subregions of central and south-central British Columbia. This subspecies is not threatened. Numbers are considered stable in most areas, but declines were reported in 1997 in the Okanagan and in both the East and West Kootenays. Generally, the highest densities occur in the southern and central parts of the province as far north as the Nechako River, and also in the northeast along the Peace River and its main tributaries around Fort St John.

Sitka Black-tailed Deer
Odocoileus hemionus sitkensis

Other Common Names: Sitka Deer.

Description

Sitka Black-tailed Deer has a darker coat than the Columbian Black-tailed Deer, and it has two white spots on the throat. The dark patch on the forehead is smaller and individuals in some populations have an obvious dark line down the nose. The metatarsal gland is 40 to 50 mm long. The Sitka Black-tailed Deer introduced to Haida Gwaii can reach high densities and individuals in some populations are smaller than those on the mainland. For other differences among the subspecies of Mule Deer, see the Description for Columbian Black-tailed Deer.

Measurements:

total length -

male: 1,567 mm (1,500-1,650) n=3

female: 1,347 mm (1,273-1,395) n=4

tail vertebrae -

male: 150 mm (150) n=3

female: 168 mm (150-195) n=4

hind foot length -

male: 435 mm (425-450) n=3

female: 407 mm (400-416) n=4

ear length -

male: 129 mm (118-139) n=3

female: 127 mm (118-140) n=4

skull length -

male: 238.0 mm (236-240) n=2

skull width -

male: 107.5 mm (107-108) n=2

Range

The distribution of Sitka Black-tailed Deer is west of the Coast Mountains beginning around Rivers Inlet and extending northward to the Portland Canal. Beyond this, the distribution becomes patchy, with some along the river valleys on the west side of Mount Robertson and in another area just south of the Yukon border at lower elevations along parts of the Alsek and Tatshenshi rivers. This subspecies was also introduced several times onto some of the islands in Haida Gwaii beginning in the late 1890s. The sources of these deer were populations on Porcher and Pitt islands. Introductions were not successful until the 1920s; but today, deer are found on most of the larger islands in this archipelago.

For those without boats, the best places to see Sitka Black-tailed Deer are probably the few accessible areas along the central and northern coast, such as around Prince Rupert, Kitimat and on Haida Gwaii.

Conservation Status

Although difficult to count, Sitka Black-tailed Deer may number more than 35,000 in the province, based on the 2011 estimate, so this subspecies is not considered threatened.

Taxonomy

The Rocky Mountain Mule Deer can be clearly separated from the other two forms in terms of external morphology, behaviour and ecology. It has been speculated that the Black-tailed Deer and Rocky Mountain Mule Deer will continue to diverge, and eventually be recognized as distinct species. Distinguishing the two coastal deer is more difficult. Mule Deer from the extreme northwest of the province on the east side of the Coast Mountains around Atlin and Tusthi lakes and Tagish, and on the north side of Graham Inlet, may be closer to Sitka Black-tailed Deer than to Rocky Mountain Mule Deer. The type specimen localities are Sioux River, South Dakota, for Rocky Mountain Mule Deer; Cape Disappointment, Washington, for Columbian Black-tailed Deer; and Sitka, Alaska, for Sitka Black-tailed Deer. Recent genetic studies of North American deer, including Mule Deer, indicate that their relationships and evolutionary histories are complex.

Traditional Aboriginal Use

Most aboriginal people in British Columbia hunted deer for food and many other uses. They used bones, such as the scapula, to make scrapers for collecting the soft inner bark or sap from trees and for scraping the hair from hides; they also fashioned bones into knives, shovels, trowels and net gauges. From limb bones they made skin scrapers, knives, adzes and chisels, needles, awls and awl holders; limb and other bones could be fashioned into necklace pieces, blanket pins, fishing points, gambling sticks, pestles for grinding pigments, and other small implements for tasks ranging from weaving to food preparation. Drummers often attached the hooves of young deer to their drum rims. In some interior and coastal groups, young people wore dew claws as rattles around the knees and ankles during puberty rites.

Like bones, antlers also provided materials for numerous tools, including digging implements, knives, tool handles, harpoon points, spear heads for fishing, hair combs, dog halter bits, headdress ornaments, wedges and clubs. Hides were valuable for clothing, footwear, drum skins and drumstick ends, pouches and bags, baby carriers, gambling mats and balls for games. The stomach and bladder made good containers for storing fat and bone marrow. Hunters used deer sinews to attach points to spear and arrow shafts, to reinforce bows, and for bowstrings and sewing threads. Interior peoples soaked skins overnight in deer brains to soften them during the tanning process.

Remarks

Odocoileus means "hollow-toothed", from the Greek *Odo* for "tooth" and *koilis* for "hollow", while *hemionus* is "mule-like" referring to the large ears of the species. Atypical coat coloration has been reported in Columbian Black-tailed Deer on rare occasions. These include albinos and individuals without any white hairs anywhere on the body. Another rare colour variant seen in adults and young consists of patches of white hair over different parts of the body.

It is normal for young of the year to be left hidden in vegetation while their mother feeds elsewhere. If you find a young deer on its own, do not assume that it has been abandoned. Too many fawns are mistakenly "rescued" by well-meaning people and end up in wildlife rescue centres. Do not handle or disturb any young deer found lying hidden.

As any gardener living in deer country knows all too well, Mule Deer can jump easily over a two-metre-high fence to feed on flowers, vegetables and fruit trees. Deer browsing creates a major problem for commercial orchards, and in the Okanagan, it has been necessary to construct stretches of tall, expensive fencing to control deer depredation. Deer browsing also causes considerable damage to forest tree seedlings. Various deterrents have been tried. Predator-odour repellents have had limited success, because like noise makers, without negative reinforcement, deer are likely to become habituated and ignore most repellents. Ungulate fencing has also been built along the Coquihalla Highway with special one-way gates (figure 57) and underpasses to allow Mule Deer and Elk to migrate relatively freely and minimize the chances of deer-automobile collisions.

Figure 57. One-way gates along the Coqhuihalla Highway in southwestern BC allow ungulates to escape from the road margins.

Selected References: Anderson and Wallmo 1984, Armleder et al. 1994, Bunnell 1990, Carr and Hughes 1993, Cowan 1936, Cowan and Guiguet 1965, Cronin 1991 and 1992, Dimmick and Pelton 1996, Geist 1990, Mackie et al. 1982, McCullough 1965, McNay and Voller 1995, Nyberg and Janz 1990, Robinson et al. 2002, Wallmo 1981, Waterhouse et al. 1993.

White-tailed Deer

Odocoileus virginianus

Other Common Names: Virginia Deer, Whitetail, White-tail Deer.

Description

The White-tailed Deer is a medium-sized deer with slender legs and a long tail covered in long hairs. The general coat colour is greyish brown in winter and reddish brown in summer. Throughout the year, the belly and chin are white, as are the backs of the lower front legs and the insides of the hind legs. There is also a white throat patch. The face is the colour of the body except for a lighter or white eye-ring, a white band at the rear edge of the black, naked nose and white inside the ears. There is a small black patch of hairs on the chin. The underside and edges of the bushy tail are white, as are the hairs on the rump under the tail. The upper side of the tail is usually the same colour as the upper body, or sometimes a darker brown. Young of the year are reddish brown with horizontal rows of white spots on the upper body.

The White-tailed Deer has at least five types of glands on its body (table 2), three of which are usually visible. There is a small antorbital gland in front of each eye and a tarsal gland on the inside of the tarsus. A small metatarsal gland (less than 40 mm long) at the lower end of the metatarsus often has white hairs in the centre. The interdigital glands between the main toes are indicated by small patches of white hair, while the nasal and poorly developed frontal glands require more invasive examination. Biologists are uncertain whether the White-tailed Deer has caudal (tail) glands. Females have four inguinal teats.

Antlers of adult males have a main beam that curves up and forward over the head. Above the bur is a moderately long brow tine branching from the main beam above the eye, followed by three to five longer, upward-pointing tines that are usually unforked (figure 42). Yearling males develop a pair of either single spikes or simple forks; the brow tine does not develop until the second year. Like Mule Deer, atypical antlers with as many as 25 points have been found in White-tailed Deer from BC. The skull of the White-tail has shallow lachrymal depressions in front of the orbits, and the vomer divides the posterior nares vertically into two sections.

Measurements:
See subspecies descriptions.

Dental Formula:

incisors: 0/3
canines: 0/1
premolars: 3/3
molars: 3/3

Identification:

The most similar species to White-tailed Deer is the Mule Deer, but there are clear external distinguishing traits. The most obvious are the rump patch and tail. In White-tailed Deer, the tail is longer and the upper (outer) surface of the tail is almost always the same brown colour as the rest of the body, not white or dark brown as in the Mule Deer. Only the white outer edges of the rump patch are visible when the tail is down in the resting position, so there is no large white rump patch always visible as in Rocky Mountain Mule Deer (figure 41). When alarmed, White-tailed Deer raise the tail and the white hairs along its edge and underneath on the rump, making the tail area appear much larger. They will also bound off waving the raised tail; called flagging, this is another unique characteristic of the species (colour photograph C-14). The ears lack the dark rim and are also proportionately smaller than those of Mule Deer. White-tailed Deer also lack the dark patch on the forehead that Mule and Fallow deer have. In summer coat, White-tailed Deer generally have a more reddish tinge than the Mule Deer. The metatarsal gland on the outside of the lower hind leg is also smaller, has white hairs in the centre, and is located toward the distal (lower) end of the metatarsal region rather than in the middle or at the proximal (upper) end, as in Mule Deer. For adult males, the non-forking tines of the antlers also separate them from Mule Deer (figure 42).

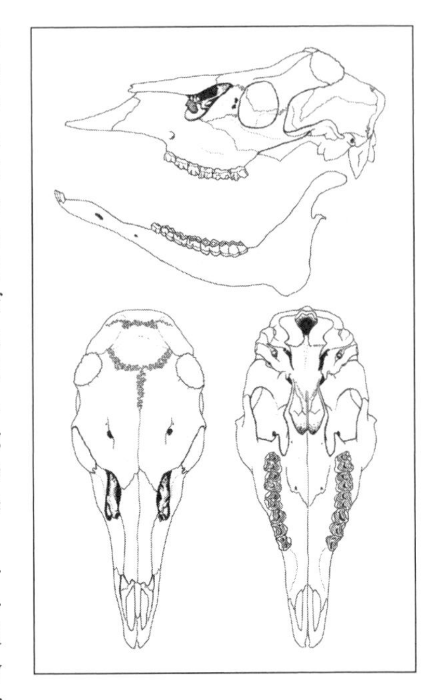

Like the Mule Deer, White-tailed Deer skulls can be separated from other similar-sized species, such as Fallow Deer, by the vomer dividing the posterior nares (figure 49). Unlike Mule Deer, the antorbital depression of the White-tailed Deer's skull is shallow (figure 50). The vomer also divides the posterior nares in Caribou, but White-tailed Deer lack upper canines and do not have the angled dorsal profile of the premaxilla.

Neither tracks nor feces can be separated easily from those of the Mule Deer, but the clearly pointed hooves of all Odocoilid deer can be distinguished from the similar-sized but blunter tracks of the Bighorn Sheep and Mountain Goat (figure 38).

Natural History

White-tailed Deer prefer heavily vegetated brush habitats, such as riparian areas, as well as extensive shrub fields. They will venture into open habitats between dusk and dawn to feed under the cover of darkness. They also occupy open forested regions in the eastern half of BC, where underbrush species of grasses, forbs and shrubs supply sufficient food and cover. In late winter, White-tailed Deer may use steep south aspects, usually between 450 and 900 metres above sea level, where snow accumulation is low and movement easy. In some areas of the province, they make spring migrations to higher elevations (up to about 1,300 metres) and spend the summer there. In the West Kootenay, they move to valleys away from the winter ranges. Movements back to winter ranges usually begin when 100 to 300 mm of snow accumulates on summer ranges. White-tailed Deer are strong swimmers when necessary, and regularly cross rivers such as the Columbia during their seasonal movements. There has been no detailed study of diets of White-tailed Deer in BC, but probably like elsewhere, the species feeds on a range of vegetation, eating mainly forbs and browse, with lesser amounts of grasses, mushrooms and the bark of young deciduous trees. In winter in the East Kootenay, Douglas-fir and Kinnikinnick were found to be important foods.

Adult male and female White-tailed Deer live apart for most of the year. The basic social unit is a family group composed of an adult female with one or two of her yearlings and young of the year. Young males leave their family when about a year old. Males are usually solitary around the rutting period but otherwise may live in small all-male groups. White-tailed Deer will often form larger temporary aggregations when attracted to choice food sources, especially in open areas.

The necks of adult male White-tailed Deer swell just before rut, and with their hard antlers, they spar with each other as well as with bushes and small trees. In serious fights between similar-sized adult males, the rivals approach and circle each other in a slight crouch, ears back and the hair on their body raised. The hairs of their tarsal glands are flared, and they seem to avoid eye-contact with each other. Then, they may suddenly lunge at each other and lock antlers, rapidly pushing and twisting with their heads close to the ground, trying to throw their opponent off balance. When one male slips or breaks away, the other tries to gore him in the rear as he flees. The winner will chase the loser for a short distance uttering a loud coughing bark, sometimes slapping his forefeet on the ground. Like most other deer, White-tailed Deer females and males without hard antlers fight with ears back and head held high, using their forefeet, either from a standing position or raised up on their hind legs (colour photograph C-17).

In British Columbia, White-tailed Deer mate in November and early December (and their courtship is quite different than that of Mule Deer). Males vigorously rub their antlers against bushes and small trees, first to remove the velvet in August, and later in the fall, perhaps as visual, auditory and olfactory signals. The noise of a male thrashing a bush travels some distance; the scratches he makes on the bark are clearly visible, and the odour he leaves behind on the vegetation, possibly from his frontal glands, is powerful. Male White-tailed Deer also scrape small patches of ground with their forefeet, and then mark the depression with urine and tarsal gland secretions using the hock-rub. They often defend these scrapes against other males. Females are sometimes attracted to them and may also urinate in them. Both sexes urinate while rubbing their tarsal glands together, but in the rut it is mainly males that do this. Both the tarsal and the metatarsal glands secrete a musky odour.

When a female starts to come into heat, she is usually courted by a single male; but sometimes, several males will pursue her, often running in large circles. A courting male approaches the female with head low like a young trying to suckle, then rushes toward her, hits the ground with his hooves, barks a few times and chases after her if she runs away. Close to estrus, the female becomes more tolerant and the male moves toward her in a crouch, sometimes bleating, and then stands alongside and guards her from other males. There are few pre-copulatory mounts, and copulation itself is quick, with the male performing a copulatory jump during ejaculation.

Gestation is 200 to 210 days. Twins are common when conditions are good, but triplets are rare. Young White-tailed Deer are born in early June and weigh 2 to 4 kg. With their spotted coats providing camouflage, they use the hider strategy in their first few weeks of life, lying hidden in thick vegetation during the day while their mothers go off to feed.

Under good conditions, female White-tailed Deer can give birth around their first birthday, but in most populations it takes place a year later. Females complete most of their body growth around 3 to 4 years of age, and males by about 4 to 6 years of age. The age of a White-tailed Deer can be determined by tooth eruption sequence up to about 3.5 years old, and by counting cementum annuli of either incisors or molars for older animals. Tooth wear does not seem to be a reliable year-specific technique. White-tailed Deer probably live for 8 to 10 years, although there are no data for populations in British Columbia. A White-tail Deer skeleton lacking a skull can be sexed by examining the pelvis. In males 2 years and older, a bump called a suspensory tuberosity projects from the anterior edge of each pelvic ilium just above their junction.

The main predators of White-tailed Deer in BC are probably Cougars and Coyotes, although Grey Wolves, Lynx and Bobcats are also known to prey on them in other areas. Bears sometimes prey on the young during the hider stage. White-tailed Deer prefer to avoid predators by hiding in dense vegetation, but if surprised, they will often snort and stamp a forefoot to give an alarm to others; they also behave this way in response to humans. When moved to flight, White-tailed Deer bound away majestically, tail raised and waving from side to side (flagging), exposing the erect white hairs underneath.

Diseases and parasites may have significant effects on White-tailed Deer populations. For example, outbreaks of epizootic haemorrhagic disease (EHD), an orbivirus closely related to Bluetongue, causes sporadic but acute mortality in White-tailed Deer. Such an EHD outbreak was reported in southeastern BC and the Okanagan Valley in the 1980s. The virus is transmitted by Biting Gnats, which also transmit the Bluetongue virus. High mortality rates during EHD outbreaks may devastate small populations. Other deer species and domestic animals generally show mild to no clinical effects from EHD infection, although deaths have been reported in Mule Deer. We do not know how this virus persists in the environment. Domestic Cattle are not carriers of EHD.

Dakota White-tailed deer
Odocoileus virginianus dacotensis

Other Common Names: Prairie White-tailed Deer.

Description
The pelage of the Dakota White-tailed Deer is generally lighter than others of this species, though it is a similar shade to that of the Northwestern subspecies. The antlers are heavy, moderately spreading with relatively short tines. It is extremely difficult to distinguish this subspecies from the Northwestern White-tailed Deer.

Measurements:
weight -
    male: 88.0 kg (86-90) n=2
total length -
    male: 1,797.0 mm (1,791-1,803) n=2
tail vertebrae -
    male: 317.5 mm (305-330) n=2
hind foot length -
    male: 501.5 mm (495-508) n=2
skull length -
    male: 290.0 mm n=1

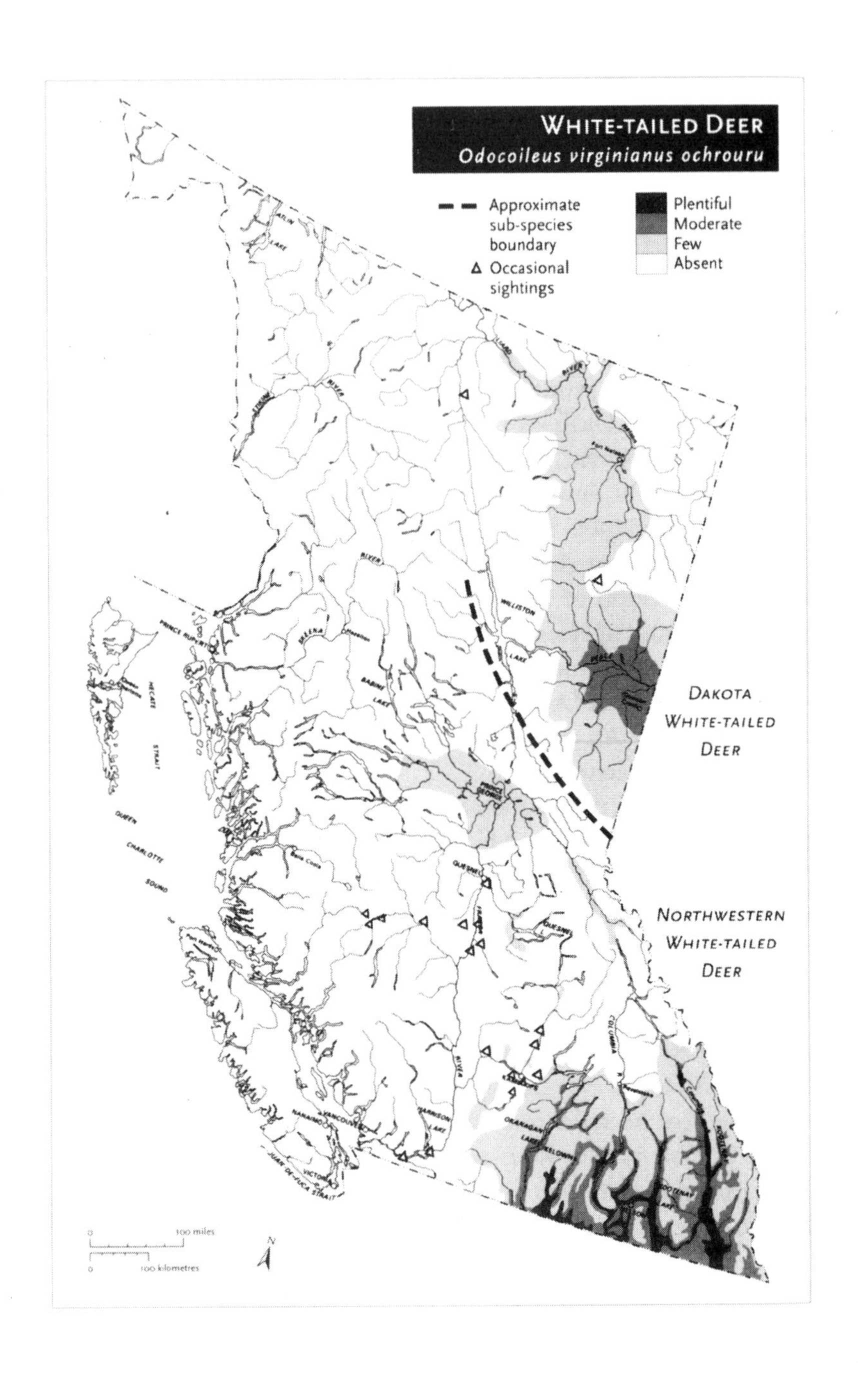
WHITE-TAILED DEER
Odocoileus virginianus ochrouru
Approximate sub-species boundary
Occasional sightings
Plentiful
Moderate
Few
Absent
DAKOTA WHITE-TAILED DEER
NORTHWESTERN WHITE-TAILED DEER
PRINCE RUPERT
HECATE STRAIT
QUEEN CHARLOTTE SOUND
WILLISTON LAKE
BABINE LAKE
PRINCE GEORGE
QUESNEL
COLUMBIA
OKANAGAN LAKE
KELOWNA
KOOTENAY LAKE
VANCOUVER
VICTORIA
HARRISON LAKE
JUAN DE FUCA STRAIT
100 miles
100 kilometres
N

Range

Although unconfirmed, it is highly probable that the White-tailed Deer in the Peace River region of northeastern British Columbia are Dakota White-tailed Deer that have spread into the province from northwestern Alberta. White-tailed Deer in this area occur at low densities, ranging west into BC from the Alberta border to the east slopes of the Rocky Mountains, and along the major rivers, such as the Fort Nelson and Peace rivers and their tributaries, as far north as the Liard River. The area between Fort St John and Dawson Creek has relatively high densities of Dakota White-tailed Deer and hence will offer the best chance of seeing this subspecies in the province.

Conservation Status

The 2011 census by the BC Wildlife Branch estimated 5,000 to 12,000 Dakota White-tailed Deer in northeastern British Columbia.

Northwestern White-tailed Deer
Odocoileus virginianus ochrourus

Other Common Names: None.

Description

The upper side of the tail of Northwestern White-tailed Deer is yellowish brown in winter. The lighter pelage colour is similar to that of the Dakota subspecies and the light underparts are more extensive. In the skull, the posterior margin of the palate is well posterior to the third upper molar. It is practically impossible to distinguish this subspecies from the Dakota White-tailed Deer in the field.

Measurements:

weight -
- male: 77.0 kg (55-112) n=3
- female: 61.6 kg (50-74) n=7

total length -
- male: ? (1,780-1,930 mm) n=?

tail vertebrae -
- male: ? (203-280 mm) n=?

hind foot length -
male: ? (500-560 mm) n=?
shoulder height -
male: ? (915-1,015 mm) n=?
skull length -
male: 270.0 mm (257-286) n=5
female: 251.3 (241-263) n=3
skull width -
male: 114.6 mm (101-123) n=5
female: 103.7 mm (103-104) n=3

Range

The Northwestern White-tailed Deer is found mainly in the southeastern part of the province, where it ranges from the Alberta border west through the Okanagan Valley to the Thompson Plateau as far as Princeton and Merritt, and north from the international border as far as Kamloops and Revelstoke to the south end of McNaughton Lake in the Columbia River valley. To the north of this area are three smaller isolated low-density populations: (1) east of Mahood Lake about 20 km north of Clearwater; (2) an arc to the west of Quesnel Lake from south of Horsefly Lake to Cariboo Lake; and (3) along the upper Fraser River from south of Tête Jaune Cache northwest to just south of the town of Dome Creek. There are also occasional sightings of individuals in this central region from Williams Lake, Quesnel, and as far west as Kleena Kleene.

A second large distribution area of what are probably Northwestern White-tailed Deer runs from Longworth in the Fraser River valley, west to Prince George, along the Nechako River to around Endako, and as far south as Punchaw Lake and north to Fort St James. The deer population in this area is sparse.

There are several reliable reports of White-tailed Deer (subspecies unknown) in the Lower Mainland, including on Herrling Island in the Fraser River east of Chilliwack, around Hope, and near Abbotsford and Matsqui. Presumably, these populations originated from individuals that escaped from the Vancouver Game Farm. The best places to look for Northwestern White-tailed Deer are low-elevation habitats and the adjacent old river benches of the Kootenay and Columbia rivers, and in the Okanagan Valley.

Conservation Status

The population estimate for Northwestern White-tailed Deer in BC in 2011 was between 62,000 and 128,000 animals. High densities of White-tailed Deer occur in low-elevation habitats associated with major river systems in the southeastern and south-central parts of the province, namely the Okanagan and Kootenays.

Taxonomy

White-tailed Deer is a widespread species in the New World, extending from northern North America, through Central America and into South America to about 15° South latitude. As many as 38 subspecies are recognized, and size variation of adults is considerable, ranging from 22 to 25 kg for the Florida Keys Deer to over 192 kg for the Northern Woodland subspecies of northeastern North America. Despite this wide variation, it is difficult to distinguish the different subspecies of White-tailed Deer in the field. Much of the variation is probably ecotypic, in response to the different environments in which the deer live. The type locality for the Northwest subspecies is the south end of Priest Lake, Idaho; and the type locality for the Dakota subspecies is White Earth River, North Dakota.

Traditional Aboriginal Use

In most cases it is difficult to identify the species of Odocoilid deer when studying artifacts, but White-tailed Deer were probably less important for aboriginal people in British Columbia than were Mule Deer, simply because they were less numerous and had a more restricted range than today. When available, they would have provided meat and hides as well as other uses similar to those from Mule Deer.

Remarks

Odocoileus means "hollow-toothed" and *virginianus* means "of Virginia". White-tailed Deer are expanding their range throughout western North America. This is also occurring in BC, as can be seen by comparing the map on page 156 with the map in *The Mammals of British Columbia* (Cowan and Guiguet 1965).

The general expansion of White-tailed Deer distribution in North America, north and westward from the southern and southeastern United States is believed to be partly the result of deforestation and agricultural modification of the environment. In some areas, there have been reports of competition and hybridization with Mule Deer. Another impact of this expansion has been the accompanying spread of the White-tailed Deer Meningeal Worm. This parasite rarely kills the White-tailed Deer, its primary host, but some researchers have suggested that it causes declines or limits the increase of other deer populations. Immature forms of the Meningeal Worm migrate into the brain or spinal cord of atypical hosts where they can cause severe, often fatal neurological symptoms. Signs include weakness, loss of co-ordination, circling, head tilt and paralysis. Fortunately, this parasite has not been reported west of the Saskatchewan-Manitoba border to date, but vigilance is important to prevent the introduction of this potentially devastating parasite into BC deer populations.

Cowan and Guiget did not include the Dakota White-tailed Deer subspecies in *The Mammals of British Columbia* (1965), nor did Nagorsen in *The Mammals of British Columbia: A Taxonomic Catalogue* (1990).

Selected References: Carr and Hughes 1993, Cowan 1936, Cowan and Guiguet 1965, Cronin 1991 and 1992, Duffy et al. 2002, Goldman and Kellogg 1940, Halls 1984, Hesselton and Hesselton 1983, Moore and Marchinton 1974, Robinson et al. 2002, Smith 1991, Wishart 1984.

Caribou

Rangifer tarandus

Other Common Names: Reindeer.

Description

The Caribou shows variation in antler and body size throughout its wide distribution across the Northern Hemisphere where it inhabits montane, boreal, subarctic and arctic habitats. In BC, Woodland Caribou is a moderately large deer with long, slender legs and large, semicircular hooves, each with a prominent dew claw just above it. The rather blunt, square nose is covered by short hair. The ears are short, broad and not pointed. Body colour varies from light brown to almost chocolate-brown in summer, and light grey or brown in winter, sometimes becoming bleached even lighter by the end of winter. The neck is usually lighter coloured than the rest of the body; females can also show some areas of lighter grey-brown just behind the shoulders. Males sometimes have a light, horizontal flank stripe above a darker belly band. The face is generally dark from the top of head to the nose, although the lower cheeks may be lighter, especially in older males. The white tail is relatively short and is surrounded by a medium-sized light or white rump patch. There is a noticeable thin band of white hairs (called socks) running above the upper edge of each main hoof. Adult males develop a mane of longer hairs along the underside of the neck as far back as the chest. During the rut, both the light neck and mane contrast strongly with the darker body.

When walking or running, Caribou make a characteristic clicking sound. This is caused by small tendons stretching over bone protuberances (sesamoid bones) in their feet. Skin glands in Caribou include antorbital, caudal, and on the hind feet, tarsal and interdigital.

The Caribou is unique among the world's deer because females regularly develop antlers. Although there is variation among the subspecies and among individuals, male Caribou antlers generally consist of a long oval-shaped main beam that rises up and back from the head before curving forward again to form a clear "C" shape when viewed from the side. The brow tine is often palmated and vertically oriented with small points along the edge, with often one brow tine larger than the other. The second tine may be branched into small points and in some individuals may be palmated. The first two tines point forward. The third, if present, is located in the middle of the main beam and projects backward; it is short, unpalmated and unbranched. The end of the main beam is usually moderately palmated, with points along the upper edge. Adult female antlers are much smaller and simpler in structure than those of adult males, being about the same size as those of one- and two-year-old males. Female antlers also lack the large, vertically palmated first tine that adult males have. Unlike other deer species in British Columbia, Caribou may grow short spiked antlers, less than 300 mm long, in their first year. The surface of the main beam of Caribou antlers is much smoother than in other deer, and the impressions of major blood vessels can often be traced along its length.

A Caribou has small upper canines that rarely project beyond the maxilla of the skull or the gum of the live animal. Upper incisors are absent and the lower incisors and canines are small and peglike. The molars have a simplified selenodont enamel pattern. A small extension of the maxilla prevents the premaxilla from meeting the nasal bone, and the anterior nasal opening is large. There is a moderately deep antorbital depression on the lachrymal bone in front of the orbit, and the pedicels are located well back on the skull so that they are partially on the frontal and parietal bones.

Measurements:
See subspecies descriptions.

Dental Formula:

incisors: 0/3
canines: 1/1
premolars: 3/3
molars: 3/3

Identification:
The Woodland Caribou is larger than the White-tailed Deer and the Mule Deer, but smaller than the Elk and the Moose. It also differs in body coloration from all other deer. Key features for identifying Caribou are the light-coloured neck and mane, the contrasting darker body, the short white tail and rump patch, and the white ring of hairs at the top of the hooves. The nose and upper lip are blunt and square.

Caribou is the only species of deer in which females and sometimes young of the year grow antlers. Relative to their body size, the antlers of adult male Woodland Caribou are the largest of all the province's deer, and their shape is unique. The long C-shaped main beam, the vertically oriented and palmated brow tines, the second tine often branched, and the palmated terminal branch are key distinguishing characteristics separating antlered male Caribou from other antlered deer. The deer with the most similar antlers is the Elk, but the lower tines of its antlers are never palmated; instead, they are sharply pointed. The surface of Elk antlers is also much rougher than that of Caribou.

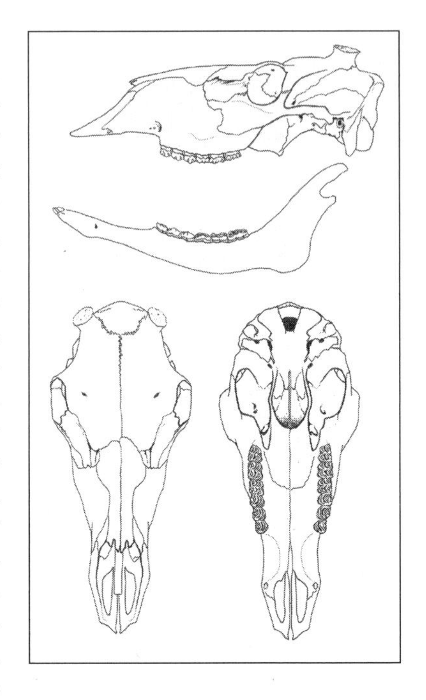

For identifying skulls, Caribou and Elk are the only deer species in the province with upper canines. But Caribou canines are much smaller than those of Elk, and are pointed and rarely project beyond the maxilla or gum. Furthermore, in the Caribou skull, the premaxilla does not touch the nasal bone and it has an angular profile when viewed from above, unlike the round shape of Elk (figure 47). The incisors of Caribou are small and round, rather than flat with sharp tips (spadelike) as in most other deer. The Caribou is only one of three species in which the vomer divides the posterior nares (figure 47), but neither of the other two species – White-tailed Deer and Mule Deer – have upper canines, and their skulls are smaller. The Caribou's anterior nasal opening is large in lateral view, so the anterior region of the skull (between the nasals and the upper tooth row) is characteristically deep. Finally, the antler pedicels are not wholly on the frontal bones but encroach onto the parietals, unlike other deer.

The Caribou's hooves are large relative to its body and they create a rounded track with the impressions of the large dew claws often seen just behind hooves, even where track impressions are not deep (figure 38). Caribou tracks may resemble those of Bison or Domestic Cattle in terms of size and general shape, but Caribou tracks are more rounded and sausage-shaped, and show their larger dew claws. The tracks of the other two show the dew claws only if the animal's foot has sunk deeply in soft ground. The clicking sound that Caribou make with their feet when they walk or trot is also unique to the species.

Natural History

Nothing is known of the natural history of the extinct Dawson's Caribou. The following information applies only to Woodland Caribou.

Biologists in British Columbia recently divided the province's Woodland Caribou into three ecotypes – Northern, Boreal and Mountain – based primarily on their winter diet and annual movement patterns. Snowfall appears to play a major role in these divisions, with Mountain Caribou having to cope with higher snow accumulations and generally more rugged mountain systems than either of the other two ecotypes. But the picture can be complicated by annual variation in snowfall.

The Northern ecotype inhabits the mountains in western and northern BC. These Caribou generally migrate twice each year, descending to low elevations in fall or early winter depending upon timing and extent of snowfall. They winter at low elevations in Lodgepole Pine or Black Spruce forests where low snowfall allows them to feed mainly on terrestrial lichens, as well as some arboreal lichens, before moving back to higher elevations in late spring and summer. They may also winter at higher elevations on open windswept ridges, where they feed on exposed terrestrial lichens before descending to low elevations again in spring to feed on early-growing plants. Females almost always move to high elevations for the calving period, while males may remain longer at low elevations. The Northern ecotype uses the Englemann Spruce-Subalpine Fir, Montane Spruce, Sub-Boreal Pine-Spruce, Spruce-Willow-Birch, and Boreal White and Black Spruce biogeoclimatic zones.

The Boreal ecotype of Woodland Caribou, included by some biologists in the Northern ecotype, occupies the flatter landscapes found in the Boreal White and Black Spruce Zone forests of northeastern BC and also feeds on terrestrial lichens in winter. In contrast to Caribou of the Northern ecotype, they are relatively sedentary and live at low densities in smaller, more dispersed groups.

The Mountain ecotype inhabits the mountainous regions of southeastern BC. In winter, because deep snow buries terrestrial foods, these Caribou feed almost entirely on arboreal lichens at high elevations. Other important foods are False Box, an evergreen shrub that is eaten in late fall and early winter in the Selkirk and Monashee mountains, and the needles of conifers that are consumed along with arboreal lichens. Animals of this ecotype often make four vertical migrations each year, moving down in early winter, back up to higher elevations in late winter, down once more to low elevations in spring and returning to high elevations for the summer. Although there is regional variation in habitat use, these migration patterns correspond to use of three biogeoclimatic zones: Interior Cedar Hemlock, Englemann Spruce-Subalpine Fir, and Alpine Tundra Parkland.

Woodland Caribou eat a wide range of forages, including grasses, forbs and browse in summer, and lichens in winter. The most commonly eaten arboreal lichens grow on coniferous trees, and include Common Witch's Hair and various species of horsehair and beard lichens. Caribou will feed on these either as they grow on standing or wind-thrown trees, or as litter blown to the ground. The ecotypes that feed primarily on terrestrial lichens in winter commonly eat lichens such as reindeer lichens, including *Cladina* and *Cetraria* species. The importance of lichens in the Caribou's winter diet is probably reflected in their tooth structure. Caribou incisors are relatively smaller and less spatulate in shape, and the enamel ridges of the cheek teeth are less convoluted than in other BC deer, probably because lichens are easily cropped and less abrasive than other forages.

Predators (Grey Wolves and Cougars), snow conditions and the availability of arboreal lichens appear to be the major determinants of habitat use by the Mountain and Northern ecotypes, especially in winter. These Caribou attempt to avoid predation by inhabiting subalpine forests well above the valley floor where wolves prefer to remain in winter. At the same time, the availability and quantity of arboreal lichens is greater in these high elevation forests. Although the snow is deeper at these high elevations, it commonly forms a crust on top that can support more weight. Snow crusting combined with the Caribou's large foot surface (created by its large hooves and dew claws) helps to keep Caribou from sinking too deeply into the snow. This allows them to travel and feed with relative ease, and as the snow pack increases, they can reach arboreal lichens growing higher up in the trees. For the other two ecotypes, the large hooves are an added advantage for pawing through snow to expose the terrestrial lichens beneath.

Woodland Caribou usually live in small groups of 4 to 6 animals, but they sometimes form larger groups of 20 to 25 in late winter at high elevations. Older males often live in all-male groups. Woodland Caribou usually show fidelity to seasonal ranges and calving areas, but sometimes may use different areas from one year to the next. Seasonal range sizes can also vary, even within a population.

Biologists are uncertain why Caribou are the only species in which females grow antlers. The most likely explanation is that they help females compete with males for food in late winter. Unlike other deer in BC, adult male Caribou shed their antlers soon after the rut, so that by November only young males, calves and females still have antlers and may keep them as late as May. (But large males in the Northern ecotype may retain their antlers until December.)

The large antlers of adult males are relatively light and are used as much for display as for fighting. During the rut, males will first use their antlers in displays and threats, but if these do not succeed, they may use them to fight. A typical fight involves shoving and wrestling, with each rival trying to gore the other; injuries and deaths can occur. Males also use their antlers to thrash bushes, perhaps as a displacement activity, or as an advertisement or signal to other males in the vicinity. Male Caribou in the rut are said to perform a behaviour called bush-gazing: standing motionless while staring fixedly at the distance. They also perform hock-urination in the rut.

The rut in BC usually begins in late September and continues through mid October when mating occurs. Woodland Caribou form harems in the mating season, with one large adult male guarding a group of adult females from other males. Males herd females using a head-up threat posture to keep them from wandering off. The harem male constantly checks each female's urine for signs of estrus by lip-curling. Caribou courtship is believed to be relatively simple compared to other ungulates.

Except for the Boreal ecotype, Woodland Caribou females in BC move and disperse to high elevations, around the treeline, to give birth, probably in an attempt to reduce predation by wolves and bears on their vulnerable young. A single young is born usually in late May or early June after a gestation period of about 228 days. Newborn Caribou weigh between 5 and 12 kg, depending on the mother's nutrition during pregnancy. Their coat is brown, often with a dark dorsal stripe, but lacks spots. Woodland Caribou females usually give birth away from their herd. Their young are followers; within a few hours after birth, the neonate is capable of following its mother.

Caribou reach sexual maturity at 16 to 28 months of age, with most females giving birth for the first time around their third birthday. Calf mortality is often high, but if they survive this vulnerable period, Caribou can live for 8 to 10 years, some for more than 15 years. A Caribou's age can be determined by tooth succession up to about 2.5 years old, and by cementum annuli counts using incisor teeth for older individuals. Adult males and females can be distinguished by antler and skull measurements, while the shorter dentary (lower jaw) length of yearling males separates them from adult females.

Predation is probably the primary cause of death. In general, the main predator of Woodland Caribou is the Grey Wolf, but the Cougar is a significant predator in the southern Selkirk Mountains. Wolverines, Grizzly Bears and Black Bears also prey on Woodland Caribou. Avalanches can be a major cause of death in some areas. As stated before, wolf predation along with lichen distribution and snow conditions seem to explain much of the Mountain and Northern ecotypes' seasonal migrations and habitat use. Not only do high-elevation forests provide abundant arboreal lichens, but the Caribou face fewer predators there. Grey Wolves prefer to remain at lower elevations in winter, where they can easily hunt deer and especially Moose.

In BC, Caribou diseases have received little research. The parasite *Besnoitia tarandi* has been found in Caribou during hunter surveys, but does not seem to cause significant problems. Bot flies and Caribou Warble Flies may be a problem for individual Caribou, but do not appear to have significant impacts on populations.

Woodland Caribou
Rangifer tarandus caribou

Other Common Names: Mountain Caribou, American Woodland Caribou, Forest Reindeer.

Description

Woodland Caribou, a medium to large deer, is the largest of all Caribou. Males can weigh as much as 300 kg and females up to 130 kg, but there are few weights available for this deer in the province. Woodland Caribou is also the darkest subspecies of any in the world. The body is usually a dark, almost chocolate brown, as is the face, belly and legs.

Woodland Caribou has the heaviest antlers of all Caribou, although they are not the longest, and their spread is often narrower. Though Woodland Caribou have the general antler plan of the species, their antlers do vary: they can be short, heavy and strongly palmated, or long with only a slightly palmated terminal section.

Measurements:

weight -

male: ? (181-272 kg) n=?

female: ? (91-145 kg) n=?

total length -

male: 2,327 mm (2,060-2,980) n=15

female: 2,007 mm (1,760-2,200) n=70

tail vertebrae -

male: 177 mm (130-220) n=10

female: 142 mm (50-220) n=51

hind foot length -

male: 548 mm (430-660) n=10

female: 528 mm (380-660) n=48

shoulder height -

male: 1,384 mm (1,245-1,580) n=14

female: 1,222 mm (1,030-1,390) n=43

ear length -

male: 149 mm (130-160) n=6

female: 143 mm (120-160) n=15

chest girth -

male: 1,447 mm (1,390-1,530) n=12

female: 1,277 mm (1,180-1,440) n=75

skull length -

male: 391.6 mm (356-471) n=9

female: 357.5 mm (331-407) n=4

skull width -

male: 150.9 mm (147-163) n=9

female: 143.0 mm (126-180) n=4

Range

In British Columbia, Woodland Caribou inhabit boreal forests and mountain regions. The Northern ecotype is spread over a large area extending north and west of Prince George that can be roughly divided into two sections. The smaller part runs from around Kleena Kleene, northwest through the Hazelton Mountains and Tweedsmuir Park as far as the Nass River to the west of Mount Weber, and also extends over parts of the western Nechako Plateau. The larger section extends across much of northern British Columbia: its northern boundary is the Yukon border; the western edge is the southern St Elias Mountains and the east slopes of the Coast Mountains to about 56° 50' latitude, through the Skeena Mountains onto the northeastern part of the Nechako Plateau east and northeast of Babine Lake. The eastern boundary is the east slopes of the Rocky Mountains. The Boreal ecotype occurs at low densities from the eastern flanks of the Rockies as far as the Alberta border, except for a small area around Fort St John where there are no Caribou.

The range of the Mountain ecotype extends from its northern limit just south of Prince George to the international border, although recent work suggests that the northern boundary may be north of Prince George at Mount Mortifee in the Misininchinka Ranges. In the Rocky Mountains, the range extends from north of Mount Robson, south to the central Rockies west of Mount Columbia. These Caribou are also found on the east side of the Fraser River through the Quesnel Highlands south of Prince George, through the Monashee Mountains to Whatshan Lake, and also through the Columbia and Purcell mountains south to about Kitchener. In the Selkirk Mountains, Mountain Caribou occur as far south as Kaslo and the east side of northern Lower Arrow Lake, then there is a break in their distribution until it begins again in the southern Purcell Mountains, where it mixes with populations from northwestern Washington and northeastern Idaho.

Accessible areas for viewing Woodland Caribou are in the northern half of the province. In winter they can been seen along the road southeast of Williston Lake. Another area with a reasonable chance of seeing them is along the Alaska Highway between Muncho Lake and Summit Lake, especially in mid to late summer. Back-country visitors may also observe Caribou in Well's Gray Provincial Park, Spatsizi Plateau Wilderness Provincial Park and Mount Revelstoke National Park.

Conservation Status

Wildlife managers recognize at least 52 discrete or semi-discrete herds of Woodland Caribou over their entire distribution in the province. They range in size from fewer than 5 individuals in the Central Rockies herd near the Alberta border to more than 3,000 in the currently stable Spatsizi herd in the north-central part of the province. In 1996, the 18 recognized herds of the deep-snow Mountain ecotype had an estimated total population of 2,300 to 2,500 animals, but by 2011, it had dropped to 1,681 animals (estimated) in 16 herds, and 5 of these had fewer than 10 members. The 31 herds of the Northern ecotype were thought to include about 17,000 Caribou, and the 6 small, scattered herds of the Boreal ecotype were estimated to total no more than about 1,000 animals. The 2011 estimate for all Woodland Caribou was estimated at around 19,600 animals. Both the Mountain and Boreal ecotypes are included in the Red List of Species at Risk in BC. The

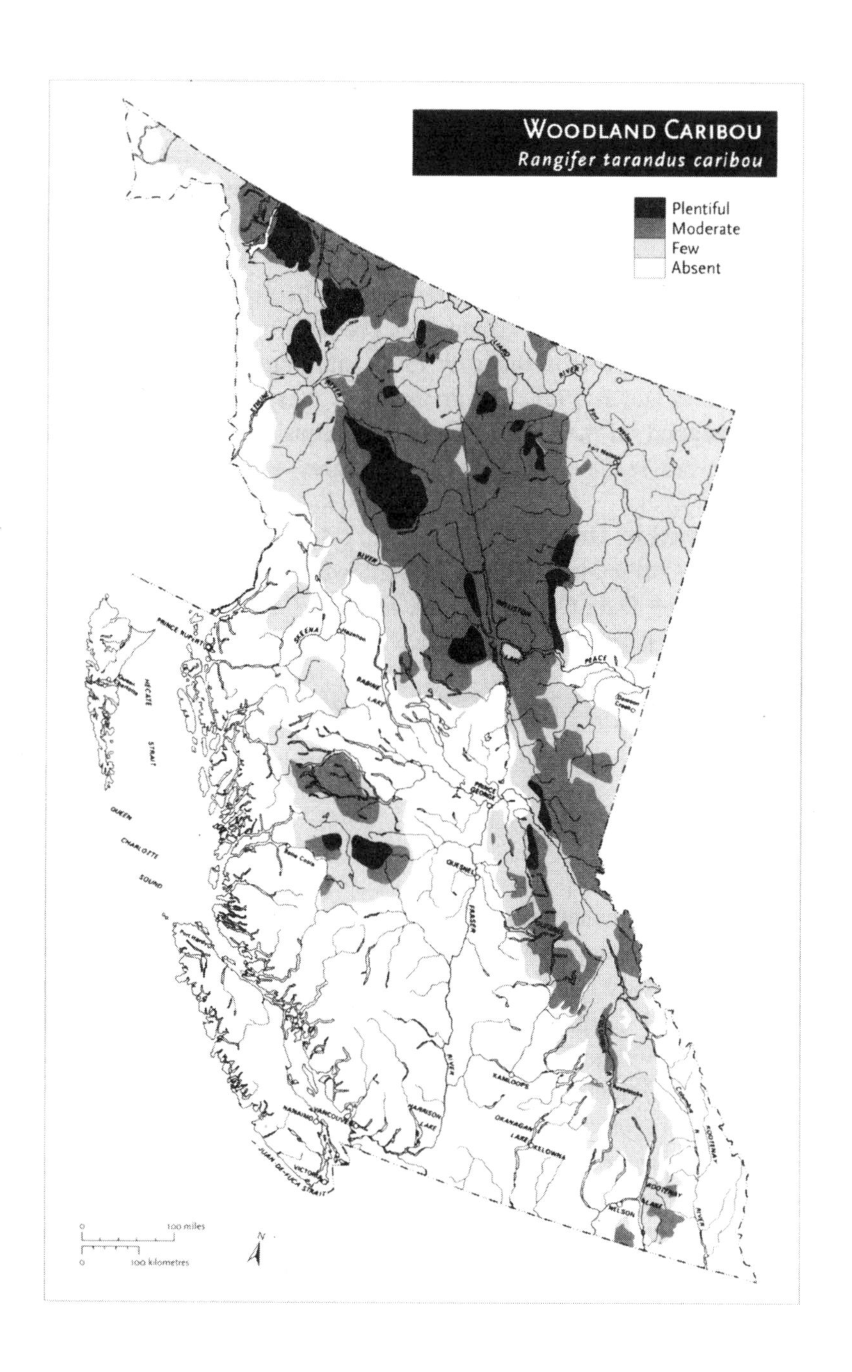
WOODLAND CARIBOU
Rangifer tarandus caribou
Plentiful
Moderate
Few
Absent
PRINCE RUPERT
PRINCE GEORGE
QUESNEL
KAMLOOPS
KELOWNA
VANCOUVER
VICTORIA
NANAIMO
NELSON
HECATE STRAIT
QUEEN CHARLOTTE SOUND
JUAN DE FUCA STRAIT
100 miles
100 kilometres

Boreal ecotype in BC is the same population that is listed as endangered in Alberta. The Northern ecotype is currently on the Blue List.

In recent times, Woodland Caribou populations in BC have undergone major fluctuations. Although major declines in southern areas must have occurred in the mid to late 1800s, the first well-recorded decline was in the 1930s and 1940s, at the same time as the range expansion of Moose and their accompanying predator, the Grey Wolf. In 1949, an aggressive wolf (and Cougar) control program was initiated, and it continued until 1962. Caribou numbers responded and increased until the late 1960s, after which they began to decline again. Current population trends vary in the recognized herds, with almost all of the Mountain ecotype in decline, while many of the northern herds appear stable. The trend is unknown for many Northern and Boreal herds.

With the troubled status of many herds, special attention to their critical habitats is a conservation priority in British Columbia. Of particular importance is maintenance of old-growth subalpine, and in some areas, low-elevation forests. The slow-growing arboreal lichens are most abundant in mature forests, and these are the essential winter food for the Mountain ecotype. Perhaps not surprisingly, conflicts arise with the forest industry because of the economic value of mature trees. Similarly, because timber harvesting can also impact terrestrial lichens as well as alter large mammal predator-prey systems, forest removal creates problems for Caribou of both the Northern and Boreal ecotypes. Several research projects have studied this problem. Most studies recommend the reservation of remaining tracts of old-growth forest within the Caribou range. Not only does logging remove the best lichen-bearing trees, but most significantly, the resulting clearcuts and young regenerating stands produce an abundance of food for Moose and deer, enabling these species to increase in number. With increasing numbers of these other more abundant ungulate species, Cougar and Grey Wolf numbers increase, and their incidental killing of Caribou is known to be excessive in such areas. Partly as a result of this habitat conflict, COSEWIC (the Committee on the Status of Endangered Wildlife in Canada) lists BC Woodland Caribou as Threatened.

Dawson's Caribou

Rangifer tarandus dawsoni

Other Common Names: Queen Charlotte Island Caribou.

Description

Dawson's Caribou is now extinct. There are only five museum specimens in the world and none of the skulls are complete. All available materials, including skins and mounts, are in poor condition. From these specimens, researchers have come up with an uncertain description. It was a small Caribou, standing about one metre at the shoulder. Its body appears to have been mouse-grey, darker on top of the head and face, with a lighter neck and belly, a short mane, and probably without the clear brown-and-white markings found in other subspecies. The antlers were poorly developed, lacking the typical Caribou palmation, and it is uncertain if females possessed them.

Measurements:

total length -
  male: 1,910 mm n=1
  female: 1,550 mm n=1
tail vertebrae -
  male: 110 mm n=1
hind foot length -
  male: 385 mm n=1
shoulder height -
  male: 1,050 mm n=1
skull length -
  male: 350 mm (est.) n=1

Range

Dawson's Caribou is known with certainty to have lived only on the plateau around Virago Sound on the northern end of Graham Island, Haida Gwaii, although an unverified skull was reported to have been found near a glacier at the mouth of the Skeena River on the mainland. Even on Graham Island, it was rarely seen, and the last three specimens were collected from a group of four in 1908. Dawson's Caribou most probably became extinct either shortly after 1910 or in the early 1920s, although there were reports of tracks as late as 1935. The ancestors of these Caribou were most likely trapped on Haida

Gwaii at the end of the last ice age when sea levels rose and cut off access to the mainland. Until recently, Haida Gwaii was called the Queen Charlotte Islands, thus the subspecies' other common name, Queen Charlotte Caribou.

Conservation Status

Dawson's Caribou is classed as Extinct by both the British Columbia Conservation Data Centre and COSEWIC (the Committee on the Status of Endangered Wildlife in Canada).

Taxonomy

Reindeer is the common name used in Europe and Asia to describe Caribou; it also refers to the domesticated form. The type specimen for Caribou comes from Lapland in Sweden; the type locality for Woodland Caribou is Quebec City, Quebec; and for Dawson's Caribou it is Virago Sound, Graham Island, Haida Gwaii. Cowan and Guiguet, in *The Mammals of British Columbia* (1965), recognized two species of Caribou in British Columbia: the Arctic Caribou (*Rangifer tarandus*), with two subspecies – Osborn's Caribou (*Rangifer tarandus osborni*) and Mountain Caribou (*Rangifer tarandus montanus*) – and the extinct Dawson's Caribou (*Rangifer dawsoni*). In his 1961 revision of *Rangifer*, Banfield recognized six subspecies in North America, only two of which occurred in BC. He included Dawson's Caribou as a subspecies of *Rangifer tarandus*, and combined Osborn's and Mountain Caribou into a single subspecies – the Woodland Caribou (*Rangifer tarandus caribou*). Ongoing genetic studies using mitochondrial and nuclear DNA from Caribou across North America will soon shed more light on the relationships between the populations and should lead to a revision of Caribou taxonomy.

Traditional Aboriginal Use

First Peoples in the north, on the coast and in the interior, valued Caribou for its meat and other body parts. They used bones such as ribs and the tibia to make many tools, including skinning knives, hide fleshers, awls and barbed spear heads for salmon fishing. They also fashioned Caribou antlers into tools for dressing skins, as well as for implements such as knives, spoons, chisels and sap scrapers, and for other items, like small boxes.

First Peoples used Caribou hides to make clothing, such as moccasins, leggings and shirts; for some winter clothes, they left the hair on the hide for insulation. Some aboriginal people used to stuff Caribou winter hair inside their moccasins to keep their feet warm. They also used Caribou skin to make fine babiche netting for snowshoes, or with the hair left on, to make warm sleeping mats. People from the central and northern interior used the whole skin from a Caribou leg as a bow-and-arrow quiver. They stored bone marrow in Caribou stomachs and made bow strings and thread from sinews.

Remarks

The generic name *Rangifer* is from the French *rangifère*, which means "reindeer". "Reindeer" originates from *reino*, the Laplander word for this animal. The specific name *tarand(r)us* is from the Latin and Greek for the fabled reindeer that could change its colour, and *Caribou* is a Micmac word meaning "shoveller", because of the way these deer paw through the snow to feed. Dawson's Caribou is named after G.M. Dawson, an early director of the Geological Survey of Canada and the first scientist to describe the animal.

Reindeer are found on a few game farms in BC. Because they are the same species as wild Caribou, hybridization is possible should any domestic forms escape from captivity.

Selected References: Banfield 1961, Bergerud 1996, Cichowski 1993, Cowan 1951, Cronin 1992, Heard and Vagt 1998, Johnson and Nagorsen 1990, Miller 1982, Rominger and Oldemeyer 1990, Seip and Cichowski 1994, Stevenson et al. 1991, Strobeck and Coffin 1996, Wittmer et al. 2005a nd 2005b, Wood 1996.

Bison
Bison bison

Other Common Names: American Bison, Buffalo.

Description
Bison is the largest of BC's ungulates. It is characterized by massive forequarters, small hindquarters, and a large head and short neck. The nose is black, square and naked, and has large nostrils. The tail is shorter than what you would expect for an animal of this size; it is covered with short hairs, except at the end where there is a tuft of long black hairs similar to that found in Domestic Cattle. Wood Bison can have longish hairs all along the tail in winter pelage. Male Bison also have a noticeable tuft of hair on the end of the penis sheath. The massive forequarters relative to the hindquarters are created by the large hump above the shoulders that is the result of the elongated spines of the thoracic vertebrae. The hump is especially large in adult males, rising abruptly about 400 mm or more from just behind the head.

The Bison's coat is dark brown, but can be bleached lighter before the end of winter, especially in older males. The head is covered by darker, almost black hair, and in both sexes the hair between the horns forms a roll or mop that can extend below and behind horns to cover

the ears in adult males. This mop of hair between the male's horns may be worn off by the end of the rut. The forehead profile of adult males is strongly convex due to its covering of long hairs. Both sexes have a noticeable beard of long hair below the chin. Generally, the beard is roughly pointed, but may develop a rounded or bulbous end in adult males. In both males and females, a fringe of long, dark hair runs along the lower margin of the neck as far as the chest to form a mane. In summer especially, the pelage of adult males is unique, with a cape of long, dense curly hairs on the anterior part of the body to just behind the shoulders, and short hair over the hindquarters. The long hair also extends down the front legs to about 10 cm above the hooves, creating the appearance of chaps or pantaloons. The cape and chaps are much less pronounced in females. The colour of the cape can range from burned sienna to reddish brown on the top and front, grading into black-brown below on the forelegs. Adult females are clearly smaller than adult males in both shoulder height and weight, being about 55 per cent of the males' weight.

Young Bison are reddish brown for the first three months of life, after which their coat colour changes to a dark brown, resembling that of the adults. White Bison are extremely rare, though not unknown, and have great significance for many First Nations of the Plains. This colour form has not been reported in BC.

In both sexes, the horns are short and curved, with sharp points. They are usually black, but can sometimes appear dark grey against the black hair of the head. With a round cross-section, the basal horn diameter is much larger in males than in females. The horn sheaths extend laterally from the head before curving sharply upward and inward. The skull is large with a broad, convex forehead, mildly protruding tubular orbits, and bone cores that project laterally in a gentle curve from the cranium. The occipital region above the condyles is flat or convex in shape.

Measurements:
See subspecies descriptions below.

Dental Formula:

incisors: 0/3
canines: 0/1
premolars: 3/3
molars: 3/3

Identification:
Bison is a massive, bulky animal, the size of Domestic Cattle or larger. Its bulky stature, together with the obvious shoulder hump and disproportionately smaller hindquarters in lateral profile, distinguish it from the almost horizontal back profile of cattle. Bison can also be identified by their dark brown shaggy coat in winter, and distinct cape of long hairs over the forequarters in summer. Other unique characteristics of Bison are the beard and the mop of hair on the top of the head between the horns. Both are especially noticeable in males, particularly on the large dark broad head of adults. These characteristics make it difficult to confuse Bison with any other ungulate in the province.

The forehead of the skull is wide between the orbits, and the somewhat tubular orbits project from the sides of the head in adult males. The pointed bone cores, present in both sexes, project laterally from behind the orbits in a gentle upward curve and usually have a strongly ridged surface in adult males. These characteristics make the skull readily distinguishable from all but Domestic Cattle. Bison skulls differ from those of Cattle in the occipital region and skull shape (figure 51). The Bison skull is much more obviously tapered from rear to front than that of Cattle. The forehead is also broader and more convex, while the orbits are more tubular and often protrude more

than those of Cattle. In dorsal view, the bone cores of Bison leave the frontal bones anterior to the para-occipital crest, whereas the cores of Cattle leave the frontal bones in line with this crest. The shape of the para-occipital crest is mildly convex in Bison, but has two bulges on either side of the centre line in Cattle (figure 51). In lateral profile, the occipital region above the condyles is concave in Cattle and convex or flat in Bison.

Tracks of Bison are similar to those of Cattle, and for most of us probably indistinguishable in the field. But in areas where there are no Cattle, the large size and rounded shape of the hoof prints are characteristic of Bison. The only other species with similarly round, semicircular-shaped hooves is Caribou, but their tracks are smaller and distinctly sausage-shaped, with clear indications of dew claws (figure 38), which in Bison are not seen unless the ground is soft and the animals have sunk in to some depth. Moose and Elk have similar-sized tracks to Bison, but Moose tracks are pointed and much narrower, and Elk tracks are less rounded and narrower (figure 34). Bison feces are almost identical to those of Cattle, varying in shape and consistency with seasonal changes in food.

Natural History

In British Columbia, Bison inhabit the Boreal Forest Region of the northeast, where the natural vegetation includes open Trembling Aspen or conifer forests and shrub lands with extensive wet and dry open meadows. Such habitats may be more typical of the Wood Bison; Plains Bison are more at home in open grasslands, typical of the rolling short-grass prairies of North America.

Bison are primarily grazers of sedges, grasses, rushes and forbs, but they also eat some browse species. A study of Wood Bison in the Slave River lowlands, Northwest Territories, identified 29 plant categories in the diet; the main plants eaten year round were Slough Sedge and reedgrass. Most Bison in other areas of North America feed mainly on grasses in summer and sedges in winter, and others feed mainly on grasses year round. It appears that the range of digestible forages available to Bison determines the array of plant species they eat. Bison appear to digest forages more efficiently than Domestic Cattle, especially plants of low quality. To forage in winter, Bison sweep their heads from side to side to clear the snow, rather than use their front feet to paw through the snow, as many ungulates do.

Bison are social animals, living in groups year round. Typically, adult males live apart from females and younger animals, so there are all-male groups and much larger maternal groups. Lone adult males of all ages are also not uncommon. Young males usually leave their maternal group when about four years old. Group size varies greatly depending on population size and density, habitat, and season. The largest groups form during the rut. When Europeans first came to North America, they reported seeing Bison herds containing thousands. Today, in large populations, maternal and rutting groups of more than 100 are common; the maximum group size is often limited by the habitat. The larger and more open the meadow, the larger the group; groups in forested habitats are small. Bison constantly vocalize in a group, uttering short grunts as they walk and feed. Grunting is typical of many large, group-living ungulates residing in open habitats, and it may help maintain group cohesion.

A notable behaviour of both sexes is wallowing (figure 58). In dry areas Bison paw shallow depressions in the ground and roll in them, creating a dusty soil that they seem to like. They make wallows throughout their range and use many of them over and over again until they are several metres in diameter. Outside the rutting season, wallowing probably has some grooming function, helping to shed the winter coat, and with the dust, helping to control biting flies.

Rutting males often utter challenging roars that can be heard from long distances (see colour photograph C-15). The roars are quite unlike the bellowing of male Domestic Cattle. Opponents approach each other head on, usually walking slowly, with tails raised (the raised tail is a sign of aggression in Bison and can be given toward humans). Two evenly matched males will stand facing each other and perform nod threats, first with the heads turned to one side, and then quickly dropping and raising their heads in unison. Nod threats are sometimes repeated for several minutes. Lateral or broadside displays are also common: the displaying male stands four to eight metres away, at a right angle to his opponent. Males often roar during displays. If the opponent moves, the displaying animal will walk in a slow, stiff-legged gait and stand in front of him again. Occasionally, both males will display laterally, standing parallel to each other. Males will also paw the ground, wallow briefly and rub nearby trees with their horns as preludes to further aggression. An animal signals submission by turning his head away, then moving away slowly.

Figure 58. An adult male Bison rolls and wallows. This behaviour compresses and erodes the ground, resulting in a noticeable depression.

If threats do not work, a fight usually results. The males face each other a few metres apart, then suddenly charge, butting heads with considerable force (colour photograph C-18). The large mat of hair on the head probably helps cushion the impact (colour photograph C-16), as does some minor pneumation of the skull. Males will also try to hook and gore each other along the sides with their short, sharp horns, occasionally with lethal results.

Bison are polygynous, and mating takes place usually between July and September. During the rut, a male seeks out a single female to court within a group. The mating unit is often referred to as a tending pair, because an adult male defends and courts only one female at a time. Male courtship is quite simple: the male stands close alongside the female while she grazes. Periodically, he checks her estrous state by testing her urine and lip-curling. When she comes into estrus, he stands at her rear, steps toward her, swings his chin toward the top of her rump, often making a soft panting sound. The female usually moves away from the male at this stage, as she does when he tries to mount her. Only when in full estrus will the female allow the male to mount her, before immediately moving away. The male must usually copulate while walking or running on his hind legs. After a successful copulation, the female often arches her back, urinates and then holds her tail horizontally for four or more hours. Most females accept only a single copulation during each mating season.

The gestation period is between 277 and 293 days, and births take place from late April to June, sometimes into July. Females usually give birth within the herd, although some may leave for a few days. A single calf is born, weighing between 14 and 18 kg; twins are extremely rare. Calves under three months of age are readily distinguished by their bright reddish-brown coat (colour photograph C-6); the reason for this distinct colour is unknown. Female Bison are protective mothers and their young are followers, staying with them until at least their first winter, although the bond between them begins to weaken noticeably in the mating season. In one instance in the Northwest Territories, herd members were seen to protect a calf from Grey Wolves, but how frequently this cooperative behaviour happens is unknown.

The growth rate of female Bison slows around three to four years of age, but in males continues until about five or six years of age. Game ranchers report that males in captivity can continue growing even longer. Most healthy female Bison bear their first young on their third birthday, but some give first birth a year sooner or later. Under favourable conditions, most females produce one young each year throughout their life; in poorer environments, they may only bear two calves every three years. Male Bison, like many wild ungulates, are sexually mature around 18 months old, but do not usually begin to be fully active in mating until at least six or seven years of age. In the field, horn shape and size can be used to determine the approximate age of a Bison up to about six or seven years in males and three to four years in females. Bison of both sexes can live more than twenty years, but the average life span is probably between ten and fifteen years.

Grey Wolves are the main predators of Bison; Grizzly Bears prey on them occasionally. Other species, such as Coyote and Wolverine, may scavenge on carcasses of animals killed by larger predators, disease or old age. The diseases important in the management of Bison include brucellosis, tuberculosis and anthrax, all probably introduced to North America by Domestic Cattle. None have been diagnosed in free-ranging Wood or Plains Bison populations in BC. Other Domestic Cattle diseases, such as respiratory viruses, Bovine Virus Diarrhoea virus and gastrointestinal parasites, have been found in captive Bison. Disease transmission to and from Domestic Cattle is potential wherever free-ranging Bison herds contact ranched Bison or Cattle herds. Another mortality factor, both historical and current, is death by drowning. This happens when Bison herds break through thin ice while crossing large rivers or lakes in late fall and early spring.

Wood Bison
Bison bison athabascae

Other Common Names: Mountain Bison, Woodland Bison.

Description
There is disagreement among biologists about the characteristics and even the existence of the Wood Bison as a separate subspecies. For example, in a recent scientific paper that re-evaluated earlier published conclusions, one of the researchers stated that the external characteristics he had used to separate the Wood Bison from the Plains Bison did not seem to hold up when he examined a larger sample size.

The Wood Bison is larger in body and skull dimensions and darker in coat colour than the Plains Bison. Its display hairs are shorter, as are the hairs on the head, beard, neck mane, chaps and penis sheath tuft; this makes the head and forequarters appear smaller than those of the Plains Bison. Conversely, the Wood Bison's tail has more hair along its length than the Plains Bison's. Its cape may also be darker.

Measurements:
weight -
  male: 943.6 kg (759-1,179) n=11
  female: 508 kg n=1
total length -
  male: 3,551 mm (3,180-3,890) n=14
  female: 2,976 mm (2,650-3,330) n=7
tail vertebrae -
  male: 496 mm (440-540) n=14
  female: 422 mm (390-480) n=7
hind foot length -
  male: 663 mm (620-710) n=14
  female: 615 mm (590-660) n=6
shoulder height -
  male: 1,822 mm (1,680-2,010) n=14
  female: 1,622 mm (1,550-1,720) n=7
skull length -
  male: 528.2 (485-555) n=19
  female: 467 mm n=1
skull width -
  male: 249.0 mm (231-268) n=7

Range

Wood Bison have been reintroduced into British Columbia, or have dispersed naturally from nearby herds living outside the province. There are currently four populations in BC. The most northern population lives mainly along the west side of the Liard River near the border of the Northwest Territories, and south to around the junction of the Liard and Crow rivers. These animals belong to the population introduced to the Nahanni Butte area of the Northwest Territories. A second BC population lives on the north side of the Liard River about 80 km west and upstream of the more northern herd. The other two populations of Wood Bison extend into BC along the Alberta border. One, along the Hay River, spends most of its time on the Alberta side near Zama Lake and only occasionally moves into BC. The other

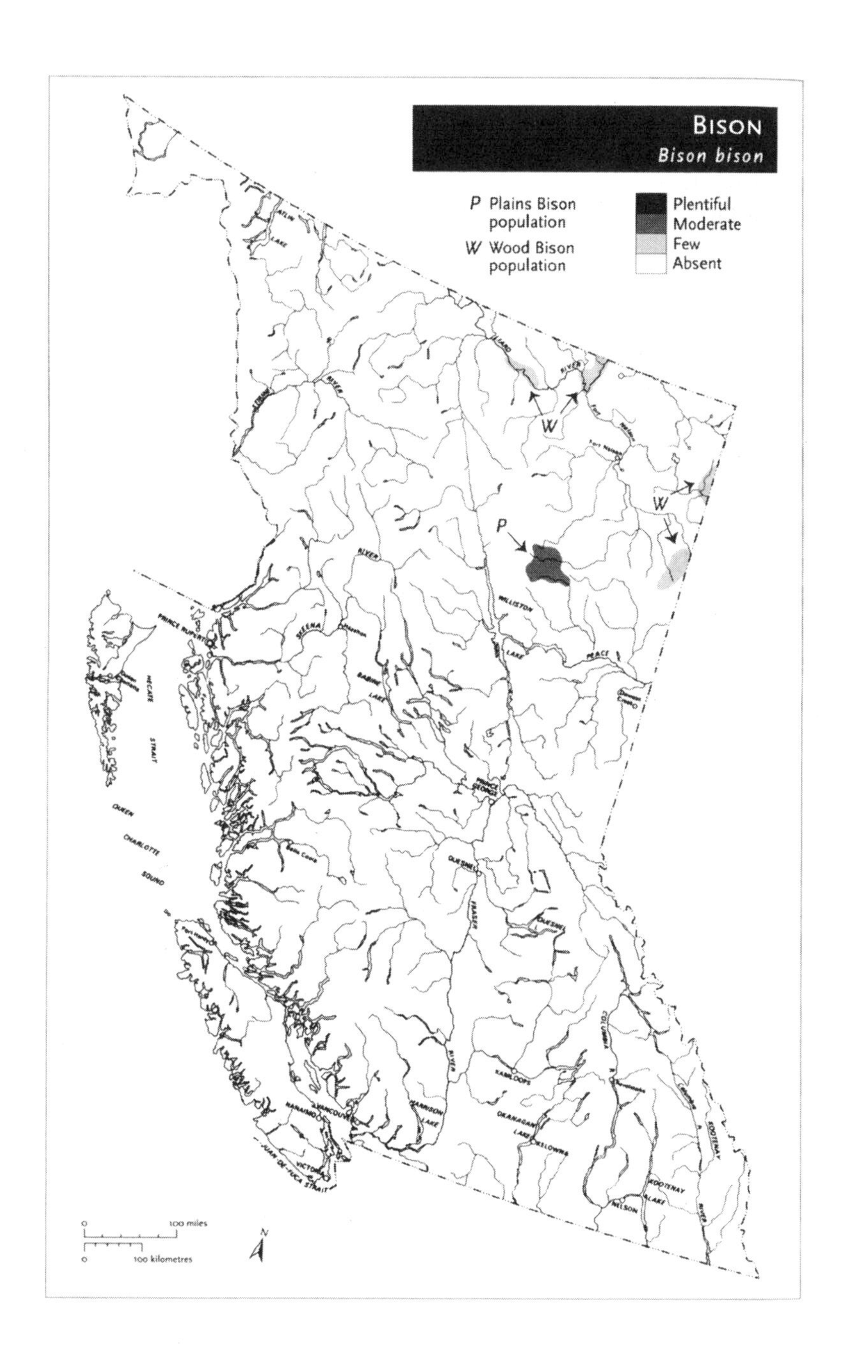
BISON
Bison bison
P Plains Bison population
W Wood Bison population
Plentiful
Moderate
Few
Absent
P
W
W
W
PRINCE RUPERT
HECATE STRAIT
QUEEN CHARLOTTE SOUND
WILLISTON LAKE
PEACE
SKEENA
BABINE LAKE
PRINCE GEORGE
QUESNEL
FRASER RIVER
COLUMBIA
KAMLOOPS
OKANAGAN LAKE
KELOWNA
HARRISON LAKE
VANCOUVER
NANAIMO
VICTORIA
JUAN DE FUCA STRAIT
KOOTENAY LAKE
NELSON
KOOTENAY RIVER
100 miles
100 kilometres

inhabits an area 80 km to the south, around the Etthithun Lake area and the headwaters of the Kahntah River, between 58° to about 57° 50' N latitude. All these transplanted Wood Bison originated from the herd in the southern part of Elk Island National Park, Alberta. Opportunities to observe Wood Bison in BC are extremely limited due to their low numbers and isolated ranges, but the area around Etthithun Lake does have vehicle access.

Conservation Status

In 2011, 400 to 600 Wood Bison were estimated in the province. This subspecies is included in BC's Red List of species at risk. COSEWIC (the Committee on the Status of Endangered Wildlife in Canada) lists Wood Bison as Threatened.

Plains Bison
Bison bison bison

Other Common Names: Prairie Bison, Plains Buffalo.

Measurements:

weight -
- male: 832.2 kg (720-998) n=5
- female: 452.5 kg (360-545) n=6

total length -
- male: 3,315 mm (3,042-3,400) n=5
- female: 2,386 mm (2,130-2,890) n=3

tail vertebrae -
- male: 399 mm (304-380) n=4
- female: 421 mm (304-508) n=3

hind foot length-
- male: 595 mm (530-680) n=4
- female: 515 mm (508-530) n=3

ear length -
- male: 150 mm n=1

shoulder height -
- male: 1,783 mm (1,670-1,890) n=7
- female: 1,520 mm (1,370-1,670) n=3

skull length -
male: 486.6 mm (450-520) n=77
female: 443.4 mm (415-468) n=25
skull width -
male: 238.8 mm (222-252) n=77
female: 207.6 mm (195-219) n=29

Range

The single free-ranging Plains Bison population in British Columbia occupies an area west of Sikanni Chief and Pink Mountain northwest of Fort St John, from just north of Trimble Lake and the Sikanni River, to the south side of the Halfway River. This population originated from a private introduction of wild Plains Bison from the northern part of Elk Island National Park in 1971, to a farm near Pink Mountain. The animals escaped shortly after arriving. These Plains Bison now occupy an area that is difficult for visitors to reach, so observing them is a problem. But several ranches in the northeastern part of the province raise Plains Bison, and while their animals are often de-horned for production purposes, they probably provide the best chances of seeing this magnificent species in BC.

Conservation Status

Plains Bison is on the provincial Red List. In 2011 between 1,100 and 1,800 were estimated in the free-ranging population. Even though released into historic Wood Bison range, this free-ranging Plains Bison herd in British Columbia is the largest herd free of problem diseases (e.g., anthrax, brucellosis) in the world, and along with the Yellowstone National Park herd in Wyoming, is considered one of the world's largest populations of this subspecies. As such, the BC herd is of major conservation significance for the species and its genotype; a management focus in BC should be to ensure it does not interbreed with animals from the Wood Bison herds.

Taxonomy

Despite many scientific papers on the subject, the subspecific distinctions between Wood and Plains Bisons are unresolved. Some researchers suggest that they are ecotypes; others maintain that they are distinct subspecies. Originally, they were separated by differences in cranial measurements, particularly those of the horn cores, but these secondary sex characteristics are notoriously variable and so are poor criteria for separating subspecies. Analysis to date shows that neither skull nor external characteristics hold consistently, nor does the analysis of blood or genetic parameters support a distinction between the two types. Having seen both types of Bison live and in photographs, and having studied and measured many skulls, I found no consistent differences between them. There was a range of phenotypic variation, particularly in hair pattern, body colour and hump shape between and within the North American Bison populations that I examined. Another researcher recently re-evaluated the external characteristics he had used to distinguish Wood Bison. He concluded that they were probably the result of the founder effect because they were based on animals from a small, isolated population of Wood Bison held in the southern part of Elk Island National Park, Alberta. Despite these findings, some researchers still maintain that there is evidence for distinguishing these two races of Bison in North America.

The type locality for Plains Bison is in New Mexico, while that of Wood Bison is a site 80 km southwest of Fort Resolution in the Northwest Territories. Another living Bison species is the Wisent or European Bison, now restricted mainly to captive or park populations in Western Europe and Russia. This species is adapted to a forested habitat and is morphologically quite different from the North American Bison.

During the early Pleistocene, probably in Asia, Bison evolved from a common ancestor with Cattle, becoming adapted to open steppe grasslands, while Cattle evolved in open forests. These early Bison were originally similar in size to the modern species, but became larger in the middle of the Pleistocene; some of these larger animals lived in BC. The horns of these large fossil forms were especially large – some spanned over two metres from tip to tip. There are several records of fossil Bison in BC. They have been found on the Saanich Peninsula on southern Vancouver Island and on some nearby islands, at Babine Lake in the central interior, and in at least four sites around Fort St John. Dates for these fossils range from around 10,000 to over 30,000 years before the present. Remains of more recent Bison, usually skulls of adult males, turn up in passes through the Rocky Mountains (e.g., around Crowsnest Pass), and occasionally in archaeological sites in the Kootenay region. In 1793, explorers reported that Bison were numerous near the confluence of the Pine and Peace rivers in northeastern BC.

Traditional Aboriginal Use

Evidence from an archaeological site north of Fort St John suggests that First Nations in British Columbia used Bison more than 10,000 years ago. This species may have been important to people living the northeast and southeast, but probably not to other First Nations in the province. Eastern groups may have used Bison meat fresh and in pemmican, and the hides for warm robes. Northern Ktunaxa (Kootenay) people were known to cross the Rockies into Alberta to hunt Bison, but this may only have been after horses were introduced to North America. The Dunne-za (Beaver) people have a story in which the Bison spirit saved them from a climatic disaster. Bison were of greater importance for aboriginal peoples on the central plains of North America.

Remarks

All free-ranging Bison in the province are the result of introductions and reintroductions either to BC or to neighbouring areas. The original animals for both the Wood and Plains Bison reintroductions into BC came from two herds in Elk Island National Park, Alberta, each originating from different sources. The Wood Bison originally came from northern Wood Buffalo National Park, and the Plains Bison from what was previously Wainwright Buffalo Park in southern Alberta, which in turn had originated from Bison from Montana.

Selected References: Berger and Cunningham 1994, Geist 1991, Geist and Karsten 1977, Harington 1996, Larter and Gates 1990, Lott 1974, Meagher 1986, Reynolds et al. 1978 and 1982, Roe 1951, Shackleton et al. 1975, van Zyll de Jong 1986.

Mountain Goat

Oreamnos americanus

Other Common Names: Rocky Mountain Goat, White Goat.

Description

The Mountain Goat is a moderate-sized ungulate with a stocky body, a noticeable hump above the shoulder seen in lateral profile, a thin neck, sharply pointed, thin black horns, and long, narrow pointed ears. The tail and lower limbs (metapodials) are short, and the hooves rather than being concave on the underside, have a thick, soft, rough-textured pad extending to the edge of the keratin hoof. The coat is completely white or yellowish white, but the nose, horns, hooves and dew claws are black. In winter, the coat consists of long guard hairs and thick underfur (wool) that are shed in June and early July for a shorter haired summer coat. Both sexes have a beard on the chin and a short mane along the underside of the neck, which is most obvious in winter. The long guard hairs on the upper two-thirds of the legs form noticeable chaps in winter. The interdigital glands are rudimentary, if present at all, and the lachrymal glands are absent, but there are black glands behind the base of each horn (the post-cornual or supra-occipital glands) that are larger in males than in females. The few weight data available for BC specimens suggest that females are 64 per cent lighter than males; but this may be misleading due to the small sample size. Based on the larger samples of body and skull dimensions, females appear on average about 12 per cent smaller than males in BC, a figure more in line with sex differences reported for Mountain Goats elsewhere in North America.

Both sexes of Mountain Goat have short horns. Adult horns are usually 200 to 280 mm long, with the longest recorded measurement of over 300 mm. The horn sheaths are shiny black and sharply pointed with a smooth surface, except at the base. They diverge slightly as they curve upward and gently backward from the head. In most males, the curve is smooth, while horns of many but not all females have a distinct bend near the tip. The horns of adult females, especially the bases, are noticeably thinner and further apart than those of adult males. The average anterior-posterior horn sheath diameter of five adult females was 33.3 mm (32–34 mm) while the average for seven adult males was 42.5 mm (36–48 mm). In rare instances, individuals may have one horn snapped off near the base, presumably the result of fighting or falling.

The skull is narrow in dorsal view and its bones are generally fragile, especially in comparison with male wild sheep. There is no lachrymal depression in front of the orbit.

Measurements:

weight -
- male: 131.1 kg (118-144) n=2
- female: 51.9 kg (43-70) n=5

total length -
- male: 1,578 mm (1,397-1,803) n=18
- female: 1,452 mm (1,321-1,575) n=24

tail vertebrae -
- male: 167 mm (140-203) n=6
- female: 128 mm (105-152) n=6

hind foot length -
- male: 350 mm (330-356) n=5
- female: 321 mm (312-343) n=6

shoulder height -
- male: 988 mm (876-1,219) n=5
- female: 899 mm (880-925) n=5

chest girth -
- male: 1,253 mm (1,067-1,422) n=12
- female: 1,060 mm (985-1,321) n=20

skull length -
- male: 267.8 mm (249-286) n=6
- female: 245.7 mm (232-267) n=7

skull width -
- male: 103.7 mm (98-112) n=6
- female: 91.8 mm (75-101) n=6

Dental Formula:
- incisors: 0/3
- canines: 0/1
- premolars: 3/3
- molars: 3/3

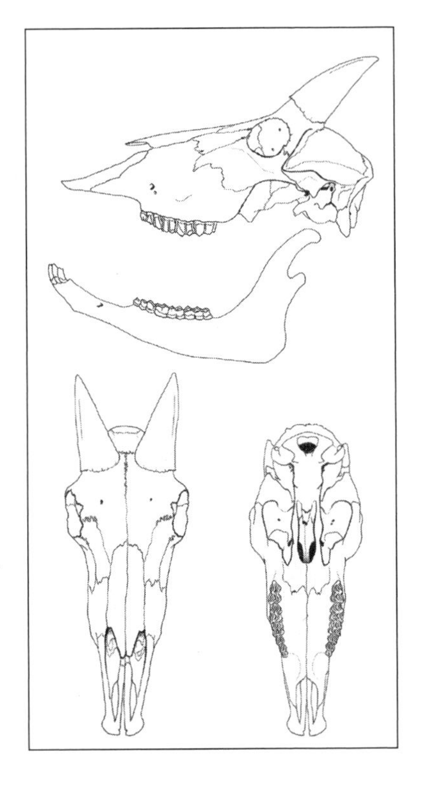

Identification:

The Mountain Goat can be distinguished from all other ungulates in British Columbia by its characteristic all-white or light-cream-coloured coat, pointed ears, short beard and short black pointed horns. The only other all-white ungulate is the Dall's Sheep, but its fur is much shorter in winter and its legs are slender and lack hair chaps. The horns of adult male Dall's Sheep are much more massive than those of Mountain Goats. The horns of female and yearling male Dall's Sheep are similar in size to those of Mountain Goats, but they are brown, ridged and blunt tipped, not black and pointed.

Some people still confuse Mountain Goats with female Bighorn Sheep where their ranges overlap, despite the different colour and the different horn and body forms. Bighorn Sheep have a brown coat, large white rump patch and a short black tail, in contrast to the all-white Mountain Goat; and the horns of female Bighorn Sheep are brown and blunt-tipped in contrast to the thin, sharply pointed black horns of the Mountain Goat.

Mountain Goat skulls are distinguished from the most similar species (e.g., female and yearling male wild sheep) by the presence of sharp, black horns, the short, straight, sharp-pointed horn cores that, like the horns, are round in cross-section, and the much narrower cranium and orbital region (figure 53). Like those of female and young male wild sheep, Mountain Goat skulls are quite fragile. The metapodials are characteristically short and robust, compared to those of wild sheep or Odocoilid deer of similar size. The soft pad of the Mountain Goat's hoof is also unique among BC's ungulates. Signs of Mountain Goats along trails are clumps of white hair they have left behind when shedding their winter coats.

Tracks of Mountain Goat are similar in size to those of Bighorn Sheep, Thinhorn Sheep, White-tailed Deer and Mule Deer (figure 38). But the impressions of Goat hooves are more rounded at the tip than those of deer, and the tracks tend to be narrower with more parallel sides than those of wild sheep (figure 38). Distinguishing fecal pellets of Mountain Goats from wild sheep and even deer, is difficult.

Natural History

Mountain Goats are remarkable climbers, able to scale extremely steep cliffs with obvious ease. Goats are found, especially in winter, in areas with steep rugged terrain such as cliffs and rock faces with ledges that they can travel along. Without doubt, they are the best climbers of all the ungulates in British Columbia, but contrary to some popular accounts, Mountain Goat hooves are not suction cups. Instead, the soft, rough-textured pads on the undersides of their hooves give them more surface friction than the hooves of other ungulates. These pads help the Goat climb on steep rock surfaces, while the hard keratin edges of its hooves help it cling to narrow rock ledges. The Mountain Goat's climbing skills are further enhanced by the flexibility of the hooves, which can splay apart when necessary to increase their surface area. Finally, its heavily built front quarters probably help the Goat haul itself upward when climbing a precipitous cliff.

Mountain Goats most frequently occupy alpine and subalpine meadows, and steep forested slopes. They are able to live on a wide variety of plant foods – grasses, forbs and much browse. The Mountain Goat's wide food habits together with an ability to tolerate deep snow for short periods probably account for its widespread distribution throughout the province. For example, in the Coast Mountains, sudden heavy snowfalls can trap Goats for several days in shallow snow wells at the bases of mature conifer trees. The Mountain Goat's ability to subsist on any available plant material means the difference between life and death.

In general, Mountain Goats, like many mountain ungulates, make seasonal migrations between high and low elevations. These movements are a response to seasonal changes in the availability of food, in snow accumulation patterns, and in microclimates. In winter, unless they can find south- or west-facing windblown slopes and ridges with sufficient food, Mountain Goats will move to lower elevations where the snow is not as deep and more food is available. In the Coast Mountains, where snow accumulations are very deep, Goats may be forced down to sea level. But in spring, when the snow begins to melt and forages grow again, the Goats move back to higher elevations. Summers are spent in alpine meadows, usually not far from rugged escape terrain, especially when predators are active in an area. Adult males in some areas have been found to spend at least part of the summer at mid elevations in forested habitat, making them difficult to observe.

In some mountain ranges, Goats make heavy use of mineral licks during early summer, even when it means travelling long distances, sometimes through heavy timber, to reach them. Males in southeastern British Columbia appear to use licks earlier in the season than females do.

Mountain Goats, primarily females and young, live in small groups of usually two to ten members. But sometimes in summer they form larger groups, most often in alpine meadows when feeding conditions are good. In some areas in BC, more than 100 animals have been counted in these mother-young aggregations in summer. Outside the mating season, adult male Goats usually live alone or occasionally in the company of two or three other males.

Both sexes of Goats are aggressive by nature and their sharp, smooth horns are dangerous weapons. If fights occur during competition, both males and females can inflict serious wounds, and even lethal injuries on rare occasions. Mountain Goats seem to avoid fighting unless they have to. Instead, males use various threat and lateral dominance displays that may draw attention to their horns or body size. Their characteristic shoulder hump probably helps increase the visual effectiveness of the lateral display. When displaying laterally, a male will lower his head, tuck his chin between his forelegs, and with arched back, walk in a stiff gait, circling and pointing his horns at his opponent. This behaviour may encourage one of the males to leave, but if not, one or both may make upward, stabbing blows with the horns, usually at the opponent's rear flank. The skin along the sides of the rump seems to be thicker, and may provide some protection against horn blows. Their characteristic aggressiveness and dangerous weapons, along with their preference for steep cliffs, may explain why their average group sizes are smaller than other mountain dwellers such as Bighorn or Thinhorn sheep. Besides a bleating sound audible at close quarters, both males and females make a variety of vocalizations, most often during aggressive encounters.

The social behaviour of Mountain Goats is poorly understood and our limited knowledge is based on only a few studies. The timing of the mating season in British Columbia varies, depending partly on latitude; in some areas, mating begins in late October, but more often, it starts in early November and continues until mid December. During the rut, males rub their horn glands on bushes; while it probably has some signal value, the function of this behaviour is unknown. During the rut, a male will also paw a depression in the ground, urinate in it, then rub his rear flanks in the wet dirt. Presumably, by covering himself with urine he increases his odour and so advertises his condition to females and rival males. In areas with dark soils, wallowing males develop dark patches along the flanks. Mountain Goats also wallow in summer; they do not urinate in the pits but roll in them for dust baths. They perform similar behaviour in snow patches in summer, possibly to reduce insect harassment or to stay cool.

A courting male will approach a female in a crouch, neck extended and nose pointed slightly upward, tongue flicking in and out. This is an extreme form of the more usual mating approach used by courting males of most other BC ungulates. The Mountain Goat male, however, must be especially careful when approaching an adult female, because she is all too ready to threaten him with her sharp horns. Once close to the female, he will nose her rear and flank, and gently kick her with a front leg. If she urinates, the male sniffs her and performs a lip-curl to test her reproductive condition. When the female is in full heat, she stands and allows the male to mount her, and she may even mount the courting male. Females are thought to remain in estrus for 48 to 72 hours.

After a gestation period of 147 to 178 days, young Goats are born in late May and early June. A female usually isolates herself in rugged terrain to give birth. Normally one offspring is born, but twins are not uncommon and triplets occur rarely. Shortly after birth, young Goats follow their mothers, who are highly protective of them. Mothers and young may form small nursery groups in spring and even larger temporary aggregations in summer. Young Mountain Goats weigh between 2 and 3 kg at birth and are all white, like adults, although some young may have a brownish dorsal stripe from the neck to the tail that disappears before it is a year old. By the end of their first autumn, young have usually developed short black horns, 25 to 65 mm long, just slightly longer than the hair on their heads. The young Goats grow rapidly, but most yearlings are smaller than adults and have a concave facial profile. Typically, two year olds can also be distinguished from adults by their smaller bodies and horns, but not in populations living in good conditions. The horns of a yearling in spring and early summer are usually shorter than its ears, but by autumn the horns are as long as the ears; those of two year olds exceed ear length. Goats are usually sexually mature at 24 to 30 months of age. Most females probably bear their first young in their third year; males do not fully participate in the rut until they are at least five or six years old.

Unlike most other ungulates in the province, Mountain Goats show limited sexual dimorphism. Horn shape is probably the best way to distinguish adult males from females at a distance; when viewing the animals close-up or examining a skull, the basal diameter of the horn is a more reliable indicator. Most adult female horns show a distinct change in curvature near the tip, whereas the curve of the male's horns is smooth. Any animal seen to squat while urinating is almost certainly a female. In short summer coat, a male's scrotum is visible.

In spring, Mountain Goats can appear to be in poor condition, because large patches of their long coat are missing or hanging loose. This is normal and is simply a sign that they are shedding their heavy winter coats. Like many other species, pregnant and lactating females usually lose their winter coats later than other age-sex classes.

Mountain Goat age can be estimated by counting horn sheath rings (annuli). For animals older than about six years, this becomes difficult because successive rings are crowded together, especially in females. The first ring, developed in the animal's first winter, is usually no more than a smooth ridge that can be difficult to distinguish. The second year's annulus is often the first easily recognizable ring. The emergence of permanent teeth and the replacement sequence of the deciduous teeth can also be used to estimate age up to three years of age; for older individuals, cementum annuli counts provide a reliable estimate.

Mountain Goats typically live around 10 or 11 years, with the oldest reported specimens being a 14-year-old male and an 18-year-old female. Not much is known about causes of mortality. Diseases (e.g., contagious ecthyma, white muscle disease) and parasites (e.g., wood ticks, lungworm) have been documented. The main predators of Mountain Goats are Cougars and Grey Wolves; occasional predators are the Bobcat, Coyote, Wolverine, Grizzly Bear and Black Bear. Like wild sheep, steep cliffs provide Mountain Goats with essential security against mammalian predators, although the cliff habitat may be less useful against Golden Eagles which sometimes prey on young Mountain Goats. Eagles have been known to knock juveniles off cliffs, then swoop down to feed on them. Harsh winters also increase Mountain Goat mortality, and are especially hard on young animals, while avalanches sometimes take a toll on all ages. When being captured and handled for transplants or research, Mountain Goats can suffer from capture myopathy (often mistakenly called white muscle disease, which is a disease of Domestic Cattle and Domestic Sheep). As a result of stress induced during capture efforts, Mountain Goats may sustain muscle damage. In extreme cases, this can cause death at the time of capture, or afterward because the animal is more susceptible to predation. In BC, selenium is a mineral often in short supply, and low selenium levels in a Goat's body may predispose it to capture myopathy.

Range

Mountain Goats occupy all the major mountain systems of mainland British Columbia, but are absent from Vancouver Island and Haida Gwaii. Except for three isolated areas along the Nelson and Chief rivers, and Boat Creek west of the Rocky Mountains, they are also absent from the Interior Plateau and Peace River Lowlands. On the west side of the province, Mountain Goats inhabit the Hazelton Mountains, and are distributed discontinuously and in varying densities along the Coast Mountains from the Yukon border to the international border. There have also been reports of Mountain Goats on some of the coastal islands (e.g., Pitt, Goat and Princess Royal islands). In the northern half of the province, Mountain Goats inhabit the Cassiar, Omineca and Skeena mountain ranges, and the Rocky Mountains as far south as Mount Garbitt, north of the Pine River. Their distribution begins again in the Rocky Mountains south of the Pine River and extends along the mountains on either side of the Rocky Mountain Trench, including through the Selkirk, Monashee and Purcell mountains as far as the international border. Mountain Goats are found generally across the southern part of the province, including the southern Okanagan.

A total of 29 translocations involving 229 Mountain Goats from BC were made between 1925 and 1996. Of these, 93 Goats were moved to locations outside the province (79 to Alberta, 10 to Colorado, 44 to Washington), where they helped repopulate regions where the species had been extirpated. In 1924, four Mountain Goats were taken from Banff, Alberta, and were introduced to an area along Shaw Creek on Vancouver Island. Given the problems created by Mountain Goats introduced to the Olympic Peninsula, Washington, it is probably fortunate that this introduction was unsuccessful on the island.

Some good locations to look for Mountain Goats are: the cliffs on the north side of the highway between Hedley and Keremeos; in Garibaldi and Cathedral provincial parks; Robson Valley in Mount Robson Park; along Highway 16 between Prince Rupert and Terrace; along Highway 37 between Stewart and Meziadin Junction; and on Rocky Ridge near Kitseguecla.

Conservation Status

Population estimates for Mountain Goats in British Columbia range from 41,000 to 66,000, making this province home to most of the Mountain Goats in Canada and over half the total world population. Most populations in BC are stable and the species is considered in no danger here. But during the winter of 1996–97, Mountain Goat numbers appeared to have declined in the East Kootenay due to severe winter conditions. Mountain Goat populations of coastal BC are in need of further study because of the importance of mature forests for winter habitat.

Taxonomy

Mountain Goats belong to the Tribe Rupicaprini (goat-antelopes) and are the only representatives of this group in North America. Other members of this tribe are the Serow and Goral that inhabit steep forested mountain slopes in southeast Asia, China and the Himalayas, and the Chamois of the major mountain systems of western and central Europe and northern Turkey. Goat-antelopes are considered to be the most primitive members of the Subfamily Caprinae.

Previously, biologists recognized several subspecies of Mountain Goat in western North America. Three occurred in British Columbia: *Oreamnos americanus americanus* ranging north from the international border through the Coast Mountains as far north as the Skeena

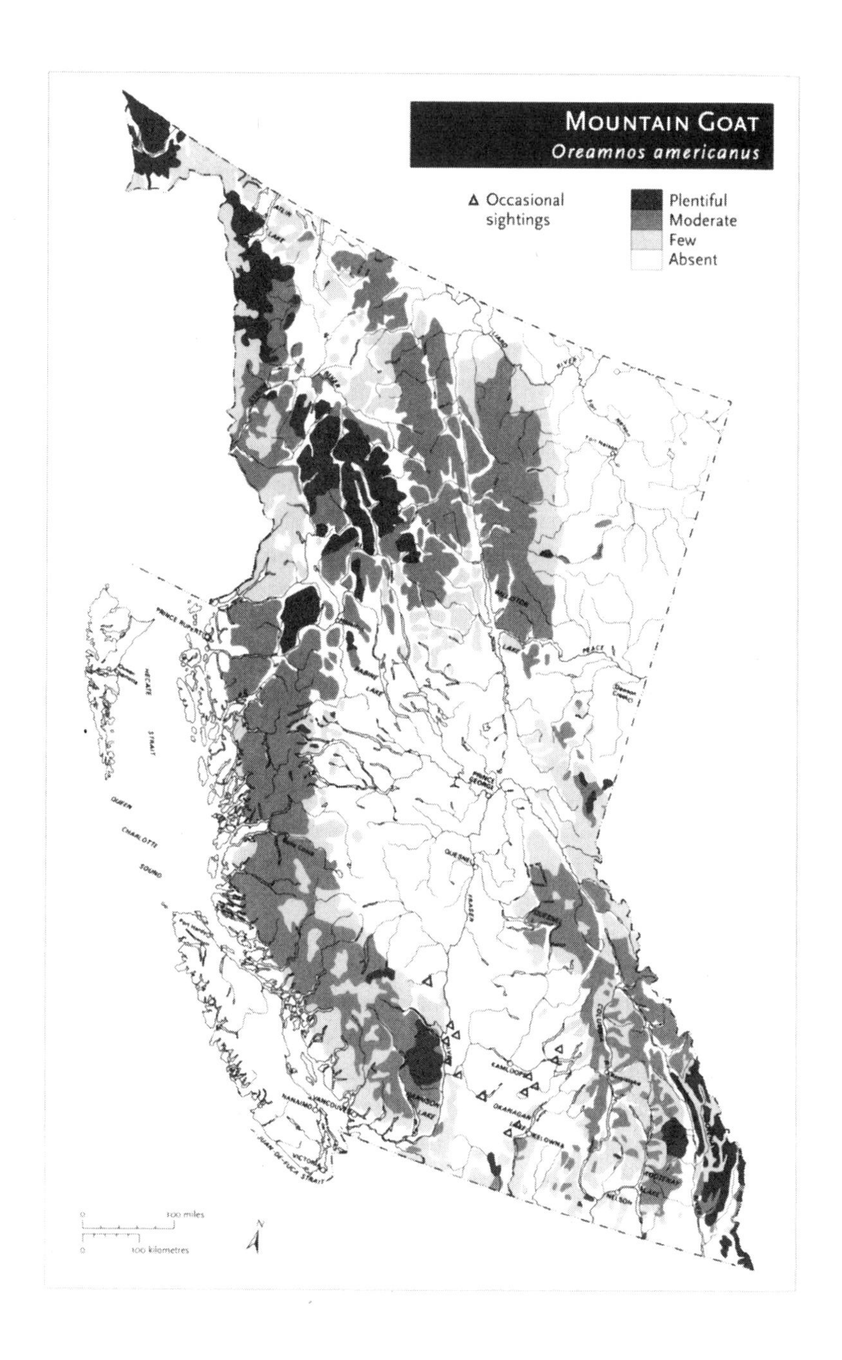
MOUNTAIN GOAT
Oreamnos americanus
Occasional sightings
Plentiful
Moderate
Few
Absent
PRINCE RUPERT
HECATE STRAIT
QUEEN CHARLOTTE SOUND
PRINCE GEORGE
QUESNEL
FRASER
KAMLOOPS
OKANAGAN
KELOWNA
VANCOUVER
NANAIMO
VICTORIA
JUAN DE FUCA STRAIT
NELSON
KOOTENAY
PEACE
LIARD
100 miles
100 kilometres
N

River, in the Selkirk and Monashee mountains to the north loop of the Fraser River, and in the Rocky Mountains from the Crowsnest Pass to the Pine River; *O. a. columbiae* in the northern half of the province north of the Peace and Skeena rivers; and *O. a. missoulae* in the Rocky Mountains south from the Crowsnest Pass. But since Cowan and McCrory's 1970 review, *Oreamnos americanus* has been considered a monotypic species (i.e., with no subspecies). The type locality for the species is uncertain, but it is probably Mount Adams in Washington.

Fossil Mountain Goat specimens are extremely rare, probably because their skulls are quite fragile. It is also rare to find modern Mountain Goat skulls in the field because they weather and disintegrate quickly, or are broken by bears and wolves feeding on them. The earliest fossil of modern Mountain Goat in North America comes from Quesnel Forks, British Columbia, in interglacial deposits prior to the last glacial period (Wisconsin). These deposits are believed to have been formed more than 90,000 years ago. In 2000, David Nagorsen found the remains of Mountain Goats on northern Vancouver Island dating from about 12,000 years ago.

Traditional Aboriginal Use

Coastal First Peoples collected the clumps of hair that caught on vegetation when Mountain Goats shed their winter coats. They separated out the fine wool underfur and wove blankets, caps and other clothing with it. They occasionally used a lichen to dye the wool yellow. The Chilkat blanket, an extremely important component of Gitxsan culture, was made of Mountain Goat wool. Coastal Peoples also fashioned Mountain Goat horns into powder containers, spoons and ladles, some of which were elaborately carved. Aboriginal peoples throughout the province occasionally ate Mountain Goat meat. Coastal and interior groups made robes from the hair-covered hides. They used the top of a Goat's skull, with the attached horns and entire skin, in ceremonial dances representing a Goat character. They also used Mountain Goat hooves in ceremonial regalia.

Remarks

The genus name *Oreamnos* is derived from the Greek *ore* for "mountain" and *amnos* for "lamb", although Mountain Goats are neither sheep nor even true goats. The specific name *americanus* refers to the species' geographic origin.

Handling Mountain Goats requires extreme caution. To avoid being injured by the sharp horns, wildlife researchers often push pieces of garden hose onto the tips when they capture an animal.

Mountain Goats are easily disturbed by low-flying aircraft, especially helicopters. Low-level flights stress the Goats, increasing the possibility of both indirect and direct mortality. Photographing or viewing Mountain Goats from low-flying aircraft, especially from helicopters, should be avoided at all costs; and heli-skiing operations need to take winter distributions of Goats into account. Helicopters should stay well away from alpine areas and cliffs used by Mountain Goats; one study recommended flying no closer than two kilometres. Researchers should also use caution when observing Mountain Goats, and especially when immobilizing and handling them. Recent research in Alberta suggests that Mountain Goats may be more susceptible to such activities than most other ungulate species, and that the impacts may be long-lasting.

Selected References: Côté 1996, Côté et al. 1998, Cowan and McCrory 1970, Festa-Bianchet and Côté 2008, Geist 1964, Hatter and Blower 1996, Hebert and Smith 1986, Poole et al. 2009, Rideout and Hoffmann 1975.

Bighorn Sheep

Ovis canadensis

Other Common Names: Mountain Sheep, Bighorn.

Description

The Bighorn Sheep is a moderately large ungulate with a stocky body, a light to dark brown coat, and a large, conspicuous white rump patch. At the end of winter, the coat colour may be bleached to a light brown. The end of the muzzle, the inside of the ears, the belly and the posterior sides of legs are white. A small, black antorbital gland is visible at the front corner of each eye; other glands include anal, caudal, inguinal and interdigital. The short tail is black and there is a black to dark brown tail stripe running anteriorly across the white rump patch. An adult male's scrotum is readily noticeable because it is white and large relative to the body size. Females have a pair of inguinal teats.

Both sexes have brown horns with transverse ridges and ripples along the surface. The horn sheaths of adult males are characteristically massive, heavily tapered from base to tip, and grow to form a curled spiral, completing a full circle or more in some older individuals. But older males may lose the tips when they become broomed. The outer edge of the horn does not form a prominent keel, but at the base, the horns of mature males are roughly triangular in cross-section. Compared to those of adult males, horns of adult females are much shorter and narrower. They have blunt tips and are oval in cross-section, curving gently up and backward from the top of the head.

Measurements:
See subspecies descriptions.

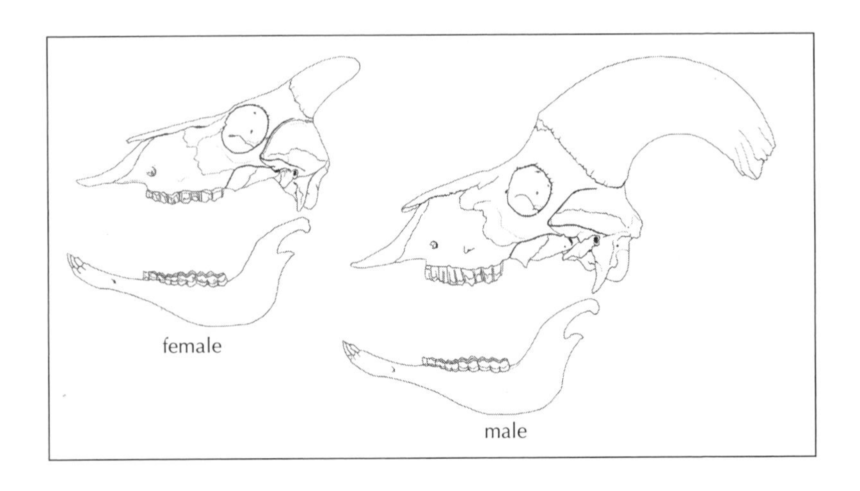

Dental Formula:
incisors: 0/3
canines: 0/1
premolars: 3/3
molars: 3/3

Identification:
The stocky brown body, large white rump patch and small black tail are diagnostic features of Bighorn Sheep. The characteristic massive brown-ridged horns of adult males are strongly curled and especially large at the base. The horns of females are much smaller, narrower, and more upright, curving only slightly up and back from the head. Although female Bighorn Sheep are sometimes confused with Mountain Goats, the two species can be readily distinguished by body coloration and horn features. Female Bighorn Sheep have a brown upper body, and even at the end of winter the large white rump patch is still visible with its short black tail, in contrast to the all-white Mountain Goat. Horns of female Bighorn Sheep are dull brownish with rounded, blunt tips, quite different from the sharp-tipped, shiny black horns of Mountain Goats.

The species most closely resembling Bighorn Sheep is the Thinhorn Sheep. Dall's Sheep is easily separated by its all-white body, while the body of Stone's Sheep is blacker or greyer than the brown of Bighorn Sheep. The lighter-coloured neck of male Stone's Sheep contrasts with the rest of the body, whereas all Bighorn Sheep, even individuals with dark brown coats, have the same-coloured neck.

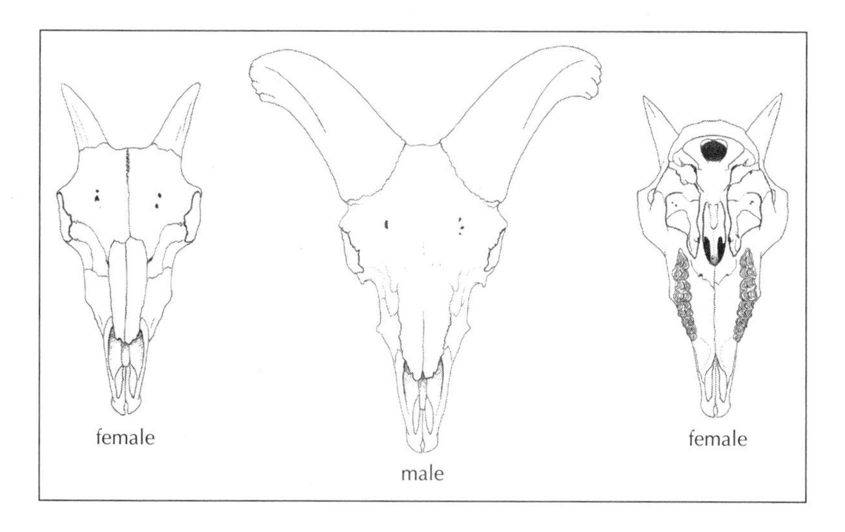

Skull features can be used to separate Bighorns from Thinhorns and from Mountain Goats. The skull of an adult Bighorn Sheep is unique because of massive bone cores and upward expansion of the cranium between the horn cores (figure 58), and the general robustness and thickness of the bones (e.g., the nasal bones). The skull of mature Bighorn Sheep can only be confused with that of an adult male Thinhorn Sheep, and is separable by the horn cores, which in Bighorn Sheep males taper less and end in relatively blunt, jagged tips (figure 58), compared to the narrower, pointed tips of Thinhorn. If horn sheaths are present, those of a Thinhorn Sheep have a more prominent outer edge (keel) and the general shape in cross-section is more clearly triangular than in Bighorn Sheep. The horn sheaths of Bighorn Sheep also tend to be pale dull brown rather than amber as in Thinhorn Sheep. Female Bighorn Sheep skulls are not easily separated from those of female Thinhorn Sheep, but can be readily distinguished from the similar-sized skull of a Mountain Goat. Horn cores of female Bighorn Sheep are oval in cross-section, being longer in the anterior-dorsal axis, rather than round as in Mountain Goat, and sheep cores have blunt, not pointed, tips. The female Bighorn Sheep skull is also relatively wider across the orbits giving the skull a somewhat triangular shape in dorsal view, compared to the narrower profile of Mountain Goat (figure 53).

Tracks of Bighorn Sheep are the same size but not as pointed as those of Mule and White-tailed deer, and the separate hoof prints are not as narrow as those of Mountain Goat (figure 34). Distinguishing fecal pellets of these four species can be difficult.

Natural History

Bighorn Sheep in British Columbia inhabit mountain ecosystems, river canyons, subalpine and alpine slopes, and high plateaus. In mountain areas, depending upon season, they range from valley floors to high alpine meadows, because sheep in many populations migrate from low winter ranges to higher summer ranges to optimize their diet. They prefer open alpine and subalpine grasslands with nearby steep, rugged terrain, such as cliffs and canyons, that they can use to escape from predators. In the south-central areas of the province, Bighorn Sheep are often associated with major river systems such as the Fraser and Chilcotin rivers, where they use the precipitous canyons flanking the river for security and the grassland benches above for feeding. Throughout their range, security habitat is extremely important, especially during the birth period when the young are easy prey for a variety of predators. Year-round, females and young generally feed closer to secure cover than do adult males, although both sexes are rarely far from it.

Bighorn Sheep favour areas with relatively low precipitation, especially low snowfall. Their predominant grazing habits mean they need grasslands. They do not depend on browse as much as other ungulate species. Unlike Mountain Goats, Bighorn Sheep have difficulty coping with snow over 300 mm deep. In mountain areas, most winter ranges are on steep south, southwest or southeast facing slopes that provide maximum exposure to sunlight. Indirect and direct solar radiation provides heat for the sheep and makes forage more available by reducing snow cover in winter and promoting early plant growth in spring. Snow-free, windblown slopes on alpine, montane and low-elevation grasslands are also used in winter because vegetation is also more readily available. Open forests, such as Ponderosa Pine forest in the Okanagan, may be used for thermal shelter in both winter and summer, but generally Bighorn Sheep prefer open habitats, rarely using heavily forested areas. They probably avoid dense forests because limited vision hampers predator detection. These different habitat requirements and tolerances probably explain the natural absence of Bighorn Sheep from heavily forested and high-snowfall ranges such as the Coastal, Purcell and Selkirk mountains, and why ill-considered introductions of sheep to such areas require winter feeding. On natural ranges, fire can play an important role in producing and maintaining Bighorn Sheep habitat such as sub-climax grasslands or open parklands below timberline. Prescribed burns have been used successfully in some areas as a habitat management tool in BC.

Figure 59. Male Bighorn and Thinhorn sheep are well known for their spectacular fights, when they hurl themselves at each other, clashing head-to-head.

Primarily a grazing species, Bighorn Sheep are opportunistic, adapting their diets to the local and seasonal changes in available plants. Besides grasses, forbs and sedges, they will also eat browse such as willows and Douglas Maple. Most browse is consumed in spring when the buds and young leaves are most nutritious. Bighorn Sheep in the southern Okanagan may browse more often throughout the year than those inhabiting the Rocky Mountains, eating shrubs such as Saskatoon, Antelope-bush and Mock Orange. Throughout their range, Bighorn Sheep seem able to survive for long periods without free-standing water if necessary, meeting much of their water requirements from succulent vegetation in summer and from snow in winter.

Like most ungulates living in open habitats, Bighorn Sheep spend their lives in groups, with adult males and females living in separate groups for most of the year. Group size varies from 2 to more than 40 sheep. The largest group reported in BC was 110 in the Okanagan, but most females and young live in groups of around 25 members.

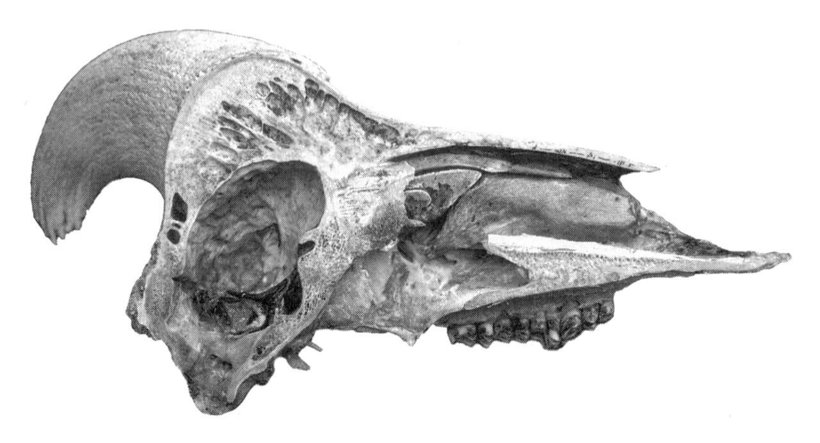

Figure 60. In the skulls of male Bighorn and Thinhorn sheep, the brain is cushioned from the tremendous impacts of their clashing fights by a space between the brain case and the outer skull. This space, called a pneumation, is created by expansions of the cornual (horn) and frontal sinuses, and has thin, flexible bone plates or struts, between the two thicker layers of bone.

Males generally live in smaller groups. The two sexes join together in the rut to form mixed groups. In many populations, groups use geographically separate areas at different seasons of the year (seasonal ranges), often returning to the same locations each year; but a few populations, such as one at Riske Creek, may stay on the same general range year-round.

Adult male Bighorn Sheep use their massive, curled horns to fight each other. Not only do they bash heads, but they rear up and run toward each other on their hind legs, hurling themselves at their opponents at the last minute (figure 59). This creates an estimated combined clash force in the order of 900 kilograms. Besides the mass of their large horns, this tremendous impact is also absorbed by an adaptation of the skull. Enlarged sinuses create a cavity within the skull above the brain (a pneumation; figure 60); the double bone plates and the strut-like connections between them are believed to cushion some of the clash force.

Besides the spectacular horn clashes, males use various displays to intimidate each other, including rearing up on their hind legs as if to clash (threat jump), and standing with head held high and staring at the opponent with the top of the horns titled toward the opponent (horn threat, usually following a clash). In extreme cases, a dominance fight can last several hours. Once a male has established his dominance, he treats the subordinate male like a female, using the same behaviour patterns he does when courting (colour photograph C-1), even including mounting. During the height of the rutting season, prolonged dominance interactions are rarely seen, because males have little time to waste in disputes over social status. Instead, they are intent in courting as many females as possible. Fights are short and sometimes vicious at this time.

The rutting season usually begins in early November and lasts until mid or late December. In some Okanagan herds, it can begin in late October. Shortly before the rut, males may gather together and interact, probably establishing dominance relationships. Most of the spectacular fights Bighorn Sheep are noted for take place at this time. The onset of the rut is signalled by males moving from female to female in a low stretch. They sniff each female's ano-genital region and she usually responds by squatting and urinating. The male will sniff the urine before raising his head to perform a lip-curl, testing her estrous condition (figure 26, colour photograph C-19). Once a dominant male finds a female in estrus, he stays with her, defending her against all other males, while also courting and copulating with her.

Courtship consists of a series of patterns that gradually decreases the distance between the male and female and then increases physical contact until the female accepts copulation. A typical sequence of courtship patterns includes the male nosing the female's flanks and rump, often while twisting his head and flicking his tongue in and out, and making a soft ululating growl. He will also gently kick or stroke the side of the female's rump with his foreleg, and later stand with his chin on her rump, pushing his chest against her rear, before finally rising up in an attempt to copulate. Females show little if any courtship toward males. Often when there are several males in a group, they will surround the courting pair, and sometimes chases will ensue. Such chases also happen early in the rut, when younger overeager males try to court females before they come into estrus.

The birth period is in early spring and coincides with new vegetation growth and warmer temperatures. After a gestation period of around 175 days, most young are born in May or June. In some Bighorn Sheep populations in the southern Okanagan and Ashnola, newborns can be seen as early as April. Females almost always produce just one young each year, but twins have been verified in some Bighorn Sheep herds in the Okanagan (e.g., the Vaseux herd). Newborns weigh from 2.8 to 5.5 kg. When about to give birth, the female usually leaves her group and moves to steep, rugged areas where her newborn will be relatively safe from most predators. Once the young is able to run fast, the pair return to the group (figure 3), usually in four to seven days. Pregnancy rates in healthy BC populations are high, up to 90 per cent. Young of the year are followers, but by mid summer, one or two adult females may be seen grazing near several young of the year, suggesting that they might be babysitting. But this may simply be because older young are more independent, and instead of tagging along near their mothers, they spend more time playing and interacting with each other, butting heads, chasing, mounting and even playing King of the Castle on available boulders.

Bighorn Sheep reach puberty as early as 18 months of age. Most females mate for the first time at about 30 months, and males much later, sometimes not until 9 or 10 years old. Young sheep do most of their growing between April and mid October, capitalizing on the new vegetation. Horn growth also takes place during this period. Females grow little after 3 to 4 years of age, while males continue to grow until at least 6 years or older. In BC, Bighorn Sheep in the Rocky Mountains are reported to be heavier and larger than those in the central part of the province, but the available data suggest no consistent differences. Each winter, most individuals lose body mass, sometimes as much as 20 per cent, or even more under harsh conditions. For adult males, their mating activities can add to the over-winter losses in weight and body condition.

Bighorn Sheep can be fairly accurately aged by counting the distinct rings (annuli) on the horns. The first horn ring can be difficult to recognize because it is often indistinct, and aging by this method works for females only up to about 5 or 6 years of age. In old males, allowance must be made where the terminal part of the horn is lost by brooming. This occurs during horn clashes between males, when the tips split and break. With a little practice, a male's age can be estimated fairly accurately in the field with the aid of binoculars or a spotting scope. Tooth succession and cementum annuli counts are other methods for aging sheep. The average longevity of adults varies between populations, but it ranges from 10 to 15 years for females, and somewhat less for males. The oldest recorded ages for individuals in the wild are from three marked Bighorn Sheep, a male and female from Banff National Park (Alberta) and a female from Stoddart Creek, BC. All three were 20 years old, and the Banff female was accompanied by a young of the year.

Mortality is usually high during the first year of life due to disease, inclement spring weather, population density, poor maternal nutrition, poor mothering, human disturbance, or predation. The impact of these factors depends in part on date of birth, range condition, population density and suitability of security cover. Coyotes are one of the most important predators of Bighorn Sheep, especially on young of the year. Mortality of first-year young can reach 80 per cent in some years where security cover is limited. Eagles also prey on young of the year. Grey Wolves, Cougars and bears take sheep of any age. Several years ago in the Junction Bighorn Sheep herd, local Cougars specialized in adult males after the rut, and together with Coyotes caused a high level of mortality.

Disease has infrequent but often devastating effects on Bighorn Sheep populations in British Columbia. More than 50 per cent of a population can be lost in one year. Less dramatic losses are more often seen where young of the year or other age classes may be selectively affected. Large-scale die-offs have been reported since the 1800s and seem to occur about every 20 years in the East Kootenay region and in some other BC herds. Such major declines are believed to have been partly responsible for major reductions in wild sheep populations throughout western North America, although overhunting around the turn of the century was the main cause of the species' drastic decline. Bighorn Sheep die-offs are still not completely understood but are thought to originate from an interaction of population and environmental stress factors along with infectious microorganisms (such as bacteria and viruses) and lungworm, which result in acute or chronic pneumonia. Mortality of Bighorn Sheep has been often associated with the presence of Domestic Sheep that can carry several infectious organisms. Die-off investigations report a range of infectious agents, including several respiratory viruses and a number of bacteria, of which some species of *Pasteurella* are the most significant, that cause rapidly developing fibrinous bronchopneumonia. Bighorns are particularly susceptible to these disease organisms, which prove fatal in most cases.

A combination of many population and environmental factors seem to play a role in a herd's overall vulnerability. These include contact with Domestic Sheep and Domestic Goats, trace mineral deficiencies, diet changes, poor nutrition, high population densities, inclement weather and other stress-producing events. Much research has been carried out into the Bighorn Sheep pneumonia syndrome. While this research has resulted in various management recommendations, such as proactive attempts to treat for lungworm and medication for bacterial pneumonia along with improving habitat quality, treatments and management to prevent die-offs have not been completely effective so far. Evidence is accumulating about the negative consequences of contact between Domestic Sheep and wild sheep.

In 1987, the introduced herd of Bighorn Sheep near Chase suffered a die-off that reduced the population from about fifty to fewer than five. Agriculture Canada diagnosed this as an outbreak of Epizootic Haemorrhagic Disease (EHD). This virus is similar to Blue Tongue and causes identical pathological lesions. This herd has also occasionally experienced less severe pneumonia-related mortality rates.

All Bighorn Sheep populations examined in BC have been infected with one or two types of protostrongylid lungworm. *Protostrongylus stilesi* and *P. rushi* are uniquely adapted to this species. Adult *P. stilesi* inhabit and lay eggs in lung tissue, and when present in large numbers can cause severe damage that often leads to secondary bacterial infections. *P. rushi* is found in the air passages and appears to be a lesser problem. Both parasites use terrestrial snails as intermediate hosts, but *P. stilesi* is also transmitted directly to the young by the mother before birth, so that infections may occur as early as six weeks of age. While lungworm in wild sheep may contribute to outbreaks of respiratory disease, they are believed to be more often responsible for reducing respiratory efficiency. This results in mortality of young animals, while for adults, heavily infected individuals are less capable of extreme effort (e.g., running to escape from predators).

Another disease of Bighorn Sheep is contagious ecthyma or soremouth, an acute to chronic eruptive dermatitis seen on the lips, nostrils, eyelids, teats or coronary bands of the hooves. It is caused by a parapox virus and is most severe in animals less than six months old, but occasionally occurs in older animals. It is transmitted by direct contact, or by contaminated feed or salt sources. It is not fatal but can reduce feeding ability and lead to debilitating secondary bacterial infections. Following an infection, most individuals are immune for life.

Bighorn Sheep occasionally show tooth problems, usually in the lower jaw. This often involves an infection with bacteria normally found in the mouth, that cause a bony growth or deformity of the jaw, called actinomycosis or lumpy jaw. Infection is believed to begin when the gum is damaged, such as when a hard, dry grass stalk or other coarse material damages the mucosa around the base of the tooth. The infection can damage the ligament around the tooth root and eventually cause the tooth to fall out. When this happens, teeth in the opposing row can become distorted if they are still erupting. Chronic jaw infections and tooth losses can reduce the animal's ability to eat because of malocclusions, and so result in poor body condition and thus hasten death. Some Bighorn Sheep may be missing one or more of the first premolars, usually in the lower jaw, but this seems to be a genetically determined condition rather than the result of actinomycosis.

Rocky Mountain Bighorn Sheep
Ovis canadensis canadensis

Other Common Names: Rocky Mountain Bighorn, Rocky Mountain Sheep.

Description

For general characteristics see the Description for Bighorn Sheep, above. The body colour of Rocky Mountain Bighorn Sheep can vary from light to dark brown. The dark colour is seen mainly in adult males and in populations from the northern and eastern parts of their range in BC. In most individuals, the dark tail stripe that runs up from the tail and across the rump patch stops just before it meets the edge of the patch, although in some of the central BC herds the stripe is often broken by the white of the rump patch. Horns of most males ten years and older are almost always broomed.

Measurements:

weight -
 male: 93.8 kg (73-120) n=58
 female: 69.6 kg (48-91) n=80
total length -
 male: 1,481 mm (1,199-1,854) n=10
 female: 1,403 mm (1,199-1,550) n=11
tail vertebrae -
 male: 113 mm (89-152-) n=7
 female: 128 mm (83-203) n=11
hind foot length -
 male: 415 mm (370-451-) n=11
 female: 401 mm (350-450) n=26
shoulder height -
 male: 1,029 mm (940-1,118) n=2
 female: 870 mm, n=1

ear length -
  male: 105 mm (102-108) n=2
  female: 106 mm (102-108) n=5
chest girth -
  male: 1,151 mm (1,110-1,200) n=6
  female: 987 mm (800-1,105) n=57
skull length -
  male: 270.8 mm (258-290) n=24
  female: 246.6 mm (235-263) n=21
skull width -
  male: 129.7 mm (126-133) n=31
  female: 116.4 mm (109-120) n=23

Range

In British Columbia, Rocky Mountain Bighorn Sheep inhabit two general areas. One is along the western slopes of the Rocky Mountains, in localized populations. Although there have been isolated sightings north of Ice Mountain (54°40'N), the most northerly part of its range begins on the BC-Alberta border in an area between the Narraway River and Jarvis Creek, extending south to about the headwaters of the Morkill River. About 170 km further south along the west slopes of the Rocky Mountains, sheep from Jasper National Park, Alberta, occasionally visit a small area in BC's Hamber Park. A further 90 km to the south, the species' most continuous distribution area in BC begins north of the Kicking Horse River at Golden, and extends southward between the East Kootenay Trench and the Continental Divide to the international border.

The second general distribution area extends from just north of Williams Lake through south-central BC to the international border. A small, isolated population may also still exist on Far Mountain at the northeast arm of Lessard Lake, east of Tweedsmuir Park. Along the western boundary of their main distribution, populations are at low densities and are somewhat discontinuously distributed on the east side of the Coastal Mountains, from northwest of Chilko Lake southeastward to the Fraser River. At the Fraser, its distribution continues south from Williams Lake along the Chilcotin and Fraser rivers, as far as the junction of Texas Creek and the Fraser River just south of Lillooet. South and east of this area, Bighorn Sheep range begins again on the west side of the northern end of Okanagan Lake around Shorts Creek near Oyama, then extends south to just north of Westbank, before recommencing on the east side of Okanagan Lake, and

Bighorn Sheep
Ovis canadensis
Occasional sightings
Plentiful
Moderate
Few
Absent
ATLIN
LAKE
RIVER
LIARD
RIVER
RIVER
WILLISTON
LAKE
PEACE
PRINCE RUPERT
SKEENA
BABINE
LAKE
HECATE
STRAIT
QUEEN
CHARLOTTE
SOUND
PRINCE GEORGE
QUESNEL
FRASER
QUESNEL
COLUMBIA
KAMLOOPS
OKANAGAN
LAKE
KELOWNA
HARRISON
LAKE
NANAIMO
VANCOUVER
VICTORIA
JUAN DE FUCA STRAIT
KOOTENAY
LAKE
NELSON
100 miles
100 kilometres
N

running south, mainly down the east side of the Okanagan Valley, as far as the international border. A small population of sheep also occur on historical range, on the east side of the Okanagan Valley, just north and west of Kaledan. These animals originated from a Vaseux Lake population that escaped from a research herd held at the Okanagan Game Farm, near Kaledan. To the west of this southerly section, relatively large numbers of Bighorn Sheep occupy an area south of the Similkameen River between Cathedral Lakes Park and the international border, and a few are also found southwest of Princeton. Bighorn Sheep are occasionally sighted in the Kettle River area near Midway, presumed to come from a population in northern Washington, some of which spend the summer in BC.

Between 1933 and 2012, there have been 58 transplants totalling 850 Bighorn Sheep in the province, many involving small numbers of animals and most within the known historical distribution of the species. The sources and destinations of these transplants often provide a complex picture, with animals being moved from several areas to establish a herd in one location, which in turn become the source for further transplants.

Here are two examples of transplants within British Columbia:

In 1966, 11 Bighorns from the Junction herd were released in an area on the north side of Kamloops Lake, between Deadman and Tranquille rivers, about 20 km west of Kamloops. Numbers increased, and in the spring of 1978, 4 young sheep crossed the North Thompson River at Heffley Creek. They continued south to occupy an area extending east of Kamloops for about 25 km on the north side of the South Thompson River. One of males disappeared and the other was accidentally killed. The 2 females remained, and a male, introduced in November, successfully mated with them. Over the next few years, small groups of sheep from Kamloops Lake, mostly males, dispersed along the same route and augmented this new population. In 1986, 22 of the 37 sheep left the area and returned to Kamloops Lake.

In 1987, another transplant introduced 6 Bighorns from the Junction population as a supplement to the herd along the South Thompson. This herd now has about 220 sheep, while the herd around Kamloops Lake numbers around 225. Bighorn Sheep are also occasionally seen on the east side of the North Thompson just south of Heffley, when animals wander up from the south and occasionally cross over from the Kamloops Lake herd, passing through the Lac du Bois area.

Bighorn Sheep have also been introduced into British Columbia from elsewhere. In 1927, 99 Rocky Mountain Bighorns were transplanted from Banff National Park, Alberta, in two separate operations: 49 went to Spences Bridge and 50 to Squilax (Chase). In 1933, 20 animals from the Squilax herd were transplanted to an area north of Squam Bay on Adams Lake, but these sheep died out or disappeared by 1964. In 1987 or 1988, 13 Bighorn from the Junction herd were released east of Louis Creek near Squam Bay. This herd still inhabits an area west of Squam Bay, although brush and forest encroachment is probably hampering any expansion in numbers. In 1970, another 12 Rocky Mountain Bighorn Sheep from Alberta were released at Spences Bridge. Today, the largest herd of Bighorns introduced from Alberta is north of Lytton, between the Fraser and the Thompson rivers northeast of Spences Bridge. From this herd a group of about 25 sheep moved along the Nicola River as far as Nicola Lake in the mid 1980s. Bighorns from another Alberta transplant inhabit an area on the south side of the South Thompson River, from Chase east to Squilax and southeast to Turtle Valley. This population suffered a major decline in 1987, so four sheep from Spences Bridge were brought in to augment the herd in 1992.

Bighorn Sheep have also naturally moved into BC from herds in the USA. The South Salmo population represents a natural northern expansion of a herd from the Hall Mountain area of Washington. This Hall Mountain herd originated from 18 sheep introduced in 1972 from Waterton Lakes National Park, Alberta, with additional animals from Thompson Falls, Montana, transplanted in 1982. Situated in the wet deep snow belt of the Interior Cedar-Hemlock biogeoclimatic zone, all three herds depend on supplemental feed in winter.

BC Bighorn Sheep herds have also been the source of transplants to locations outside the province, the last of which occurred in 2000. A total of 589 Bighorns have been transplanted in 32 operations from BC to sites in California, Colorado, Idaho, Nevada, North Dakota, Oregon, Utah and Washington.

Some of the most promising places to see Rocky Mountain Bighorn Sheep in their native range are in the East Kootenay region south from Radium Hotsprings (Kootenay National Park) and along the west-facing slopes and benches of the Columbia River valley. Bighorns are also frequently observed in four other areas: along the Fraser and Chilcotin river canyons near their confluence; on the north side of Kamloops Lake and the South Thompson River from Highway 5 east to Monte Creek; on the east side of Vaseux Lake; and around Ashnola and Ewart creeks east of Cathedral Park. Sheep in the introduced populations are often seen along the north side of the confluence of the Fraser and Thompson rivers to Spences Bridge, at Kamloops on

Highway 5 at the corner of the North and South Thompson rivers, along Highway 1 above Chase, and on the north side of Highway 3 just west of the summit of the Salmo-Creston highway.

Conservation Status

The 2012 population estimates for Rocky Mountain Bighorn Sheep in British Columbia are between 5,900 and 7,200. Most herds range in size from fewer than 10 to more than 400 individuals and are stable or declining, although some are increasing. Between 1993 and 1995, reported declines were associated with heavy lungworm infestations. In 1996–97, the southern interior experienced a severe winter with resulting heavy losses to ungulate populations, including Bighorn Sheep. In addition, major die-offs of Rocky Mountain Bighorn Sheep, which have been recorded as recently as the 1980s in BC (see Natural History above), are always of concern. Some disease-related declines can be prevented by ensuring that Domestic Sheep are kept separate from wild sheep.

Currently, Rocky Mountain Bighorns in BC are threatened by loss of winter ranges to developments on private lands, displacement from certain lambing grounds by rock climbers in spring (e.g., in the southern Okanagan), forage competition with livestock, overly optimistic hunting limits and highway mortalities. In many areas, their habitat is deteriorating due to invasive weeds, the elimination of natural fires and, possibly, overgrazing by the sheep themselves. These threats, together with its relatively low numbers and small populations, mean that the Rocky Mountain Bighorn Sheep is included in the Blue List of species at risk in BC. If these problems can be mitigated by provincial wildlife managers, the future of Bighorn Sheep should be secure. Management prescriptions include range management actions such as improving livestock grazing systems, land acquisition, access planning and management, and controlled burns to stop forest encroachment and increase grassland diversity. Additional protocols that may have some value include lungworm treatment and provision of trace minerals, but only if their limitations are evaluated.

Taxonomy

Prior to 2001, two subspecies of Bighorn Sheep were recognized in British Columbia: California Bighorn Sheep (*Ovis canadensis californiana*) and Rocky Mountain Bighorn Sheep (*O.c. canadensis*). Recent DNA and morphometric studies found no reason for separating California from Rocky Mountain Bighorn Sheep, except for those living in the Sierra Nevada of California. The California Bighorn was found naturally in south-central BC, south from just north of Williams Lake to the international border. The natural range of Rocky Mountain Bighorn in BC was formerly recognized as being along the western slopes of the Rocky Mountains. There are significant ecological and environmental differences between these two areas, so it is possible that, in future, they might be recognized as ecotypes adapted to their different environments.

Fossil remains of Bighorn Sheep are relatively rare in North America, but a surprising number have been found in BC. The best specimens have come from around Kamloops and along the Parsnip River, near Finlay Forks. This northeastern area is no longer inhabited by either Bighorn or Thinhorn sheep. Most of the BC fossils date from between 5,000 and 10,000 years ago. The type locality for Rocky Mountain Bighorn Sheep is Exshaw, Alberta.

Traditional Aboriginal Use

Bighorn Sheep meat and hides were used by First Nations living within the sheep's range. Sheep bones were used for drill handles, combs and long knives. Horns were fashioned into bowls and ladles by interior and coastal groups, the coastal people presumably obtaining them through trade. The horns were probably first heated in water before being bent and moulded, followed by trimming to the desired shape. The resulting utensils were sometimes embellished with carvings and then polished to become superb works of art.

Remarks

Ovis is Latin for sheep, and the specific name *canadensis* refers to the type specimen of Bighorn Sheep coming from a population in the Canadian Rocky Mountains that exists to this day north of Exshaw, Alberta.

Hunters around the world consider Bighorn Sheep, especially adult males from the Rocky Mountains, one of the most highly prized big-game animals in North America, if not the world. Consequently, much effort is spent managing populations and their habitats in

BC and elsewhere throughout the species' range. Even so, a study in Alberta provided evidence that selective trophy hunting of Bighorn males could have long-term deleterious effects.

Selected References: Coltman et al. 2003, Cowan 1940, Eccles and Shackleton 1979, Geist 1971, Hatter and Blower 1996, Hebert and Harrison 1988, Holm and Reid 1975, Lawrence et al. 2010, Portier et al. 1998, Schwantje 1988, Shackleton 1985, Shackleton et al. 1999, Shackleton and Hutton 1971, Weyhausen and Ramey 2000, Wikeem and Pitt 1992.

Thinhorn Sheep

Ovis dalli

Other Common Names: Thinhorn.

Description

The Thinhorn Sheep is a medium-sized ungulate with a stocky body, slender legs, short ears and short tail. The body colour ranges from the all-white Dall's Sheep to the dark blackish-brown forms of Stone's Sheep, with a general cline of all-white individuals in the northwest to darker ones further south and east in the species' range. Coat colour is the primary character used to distinguish the two subspecies.

Both sexes have horns, which are generally light brown or amber, with rings and ridges over their surface. Adult males have much larger horns than females, and they are roughly triangular in cross-section with an obvious keel on the upper outer edge. Like those of most male

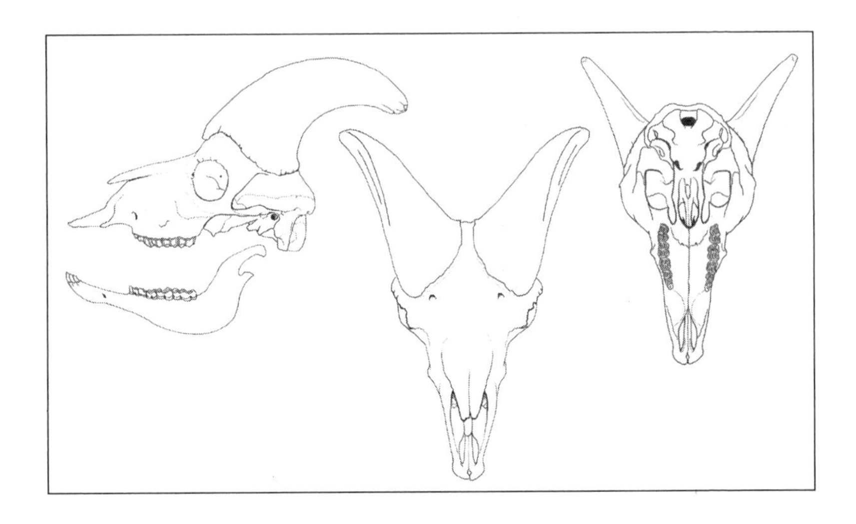

wild sheep, the horn sheaths are sharply tapered and grow in a spiral out from the head so that in mature individuals they complete a full circle alongside the face. Most adult males – even the old ones – have a wide horn span and unbroomed horn tips. The female's horns are short, gently curved up and back from the head, and elliptical in cross-section.

Measurements:
See subspecies descriptions.

Dental Formula:

incisors: 0/3
canines: 0/1
premolars: 3/3
molars: 3/3

Identification:
The only other BC ungulate with the all-white coloration of Dall's Sheep is the Mountain Goat. Both sexes of Dall's Sheep are readily distinguished from Mountain Goat by their lack of a beard and more slender legs, and in winter, by their shorter body hair. Male Dall's Sheep have much larger and massive horns than Mountain Goats; they grow in a spiral and range in colour from light amber to dark brown, rather than black like a Mountain Goat. Female Dall's Sheep horns are similar in size to those of Mountain Goats, but have blunt (not sharp) tips, are elliptical (not round) in cross-section, have a

ridged (not smooth) surface and are amber or dark brown (not black). Despite their colour variations, Stone's Sheep never seem to show the dull, medium-brown coat colour typical of either Rocky Mountain or California Bighorn Sheep. Also, compared to Bighorn Sheep, Dall's and Stone's Sheep horns are more amber coloured, and in males have a more prominent outer keel and are less prone to brooming. See also the Identification section for Bighorn Sheep.

Natural History

Thinhorn Sheep inhabit open mountain slopes up to the alpine zone and eat grass and low shrubs. Most aspects of their basic biology, ecology and social behaviour are similar to those described for Bighorn Sheep. Some data suggest that Stone's Sheep may live in smaller groups than Bighorn Sheep.

Mating usually begins in mid to late November and runs to late December in BC. Gestation is about 175 days, and females invariably bear a single young during May or June. The few recorded birth weights range from 3 to 4 kg. Like Bighorn Sheep, female Thinhorns give birth away from their group and when they rejoin it, their young use the follower strategy. Sexual maturity is attained at 18 to 30 months of age, depending upon population conditions, but males fully participate in mating only when much older. The age of a Thinhorn Sheep can be determined by tooth succession, cementum annuli counts or the distinct rings on their horns, with the same limitations as described for Bighorn Sheep. The average longevity of Thinhorn Sheep is probably around ten years for both sexes, but older animals are not uncommon.

The main predators of Thinhorn Sheep are Grey Wolves, Grizzly Bears, and sometimes Wolverines and Lynx. Golden Eagles may also prey on young sheep in the first few weeks of life. The sexes differ in their response to predators in spring, when females with young remain closer to security than do large males. Like Bighorns, Thinhorn Sheep frequently carry one or two types of protostrongylid lungworm. Although little monitoring has been done, Thinhorn populations do not appear to suffer the die-offs experienced by Bighorn Sheep herds, possibly because they rarely reach the high densities that some Bighorn populations do and because their ranges are not used by domestic livestock. The tooth problems described in Bighorns may be more common in Thinhorn Sheep.

Dall's Sheep
Ovis dalli dalli

Other Common Names: White Sheep, Dall Sheep.

Description

Dall's Sheep is characterized by its overall white or creamy white coat and tail. This all-white pelage distinguishes it from Stone's Sheep with its grey to black body, white rump patch and black tail. But some populations (mainly in the Yukon) have intermediate forms between Dall's and Stone's Sheep. Some individuals may be all white with only a black tail; others can show light grey patches, often on the middle of the back.

Measurements:

weight -

male: 84.1 kg (70-103) n=5
female: 48.8 kg (46-51) n=8

total length -
    male: 1,552 mm (1,346-1,740) n=11
    female: 1,372 mm, n=1
tail vertebrae -
    male: 94 mm (70-121) n=9
    female: 85 mm (70-90) n=10
hind foot length -
    male: 402 mm (370-432) n=10
    female: 338 mm (350-390) n=10
shoulder height -
    male: 988 mm (927-1090) n=9
ear length -
    male: 89 mm (85-92) n=6
    female: 87 mm (80-90) n=9
chest girth -
    male: 1,159 mm (1,100-1,240) n=5
    female: 1,085 mm (1,050-1,120) n=2

skull length -
    male: 249.0 mm (235-256) n=16
    female: 228.0 mm (220-238) n=4
skull width -
    male: 122 mm (108-116) n=7
    female: 115 mm (112-117) n=3

Range

In British Columbia, Dall's Sheep is restricted to the southern extension of the St Elias Range in the Haines Triangle west of Bennett Lake in the extreme northwestern portion of the province. Outside BC, the subspecies inhabits the Mackenzie Mountains along the border between the Yukon and the Northwest Territories, and various mountain ranges in the Yukon and Alaska.

It is not easy to see Dall's Sheep in the province. Visitors prepared to travel to the northwestern corner of BC may have better luck going a little further into the Yukon and visiting Sheep Mountain in southern Kluane National Park Reserve, southwest of Haines Junction.

Conservation Status

The number of Dall's Sheep was estimated at between 400 and 600 in British Columbia in 2011, only a fraction of the total world population of around 100,000. Because of their low number in BC, Dall's Sheep is included on the province's Blue List of species at risk.

Stone's Sheep
Ovis dalli stonei

Other Common Names: Black Sheep, Stone Sheep.

Description

Stone's Sheep varies in coat colour from light grey to a grizzled grey-brown to almost black; it tends to be darker in the southern and eastern parts of its distribution, and lighter in the north and northwest. A population of Stone's Sheep may contain almost the entire range of coat variation, although one colour type generally predominates. The

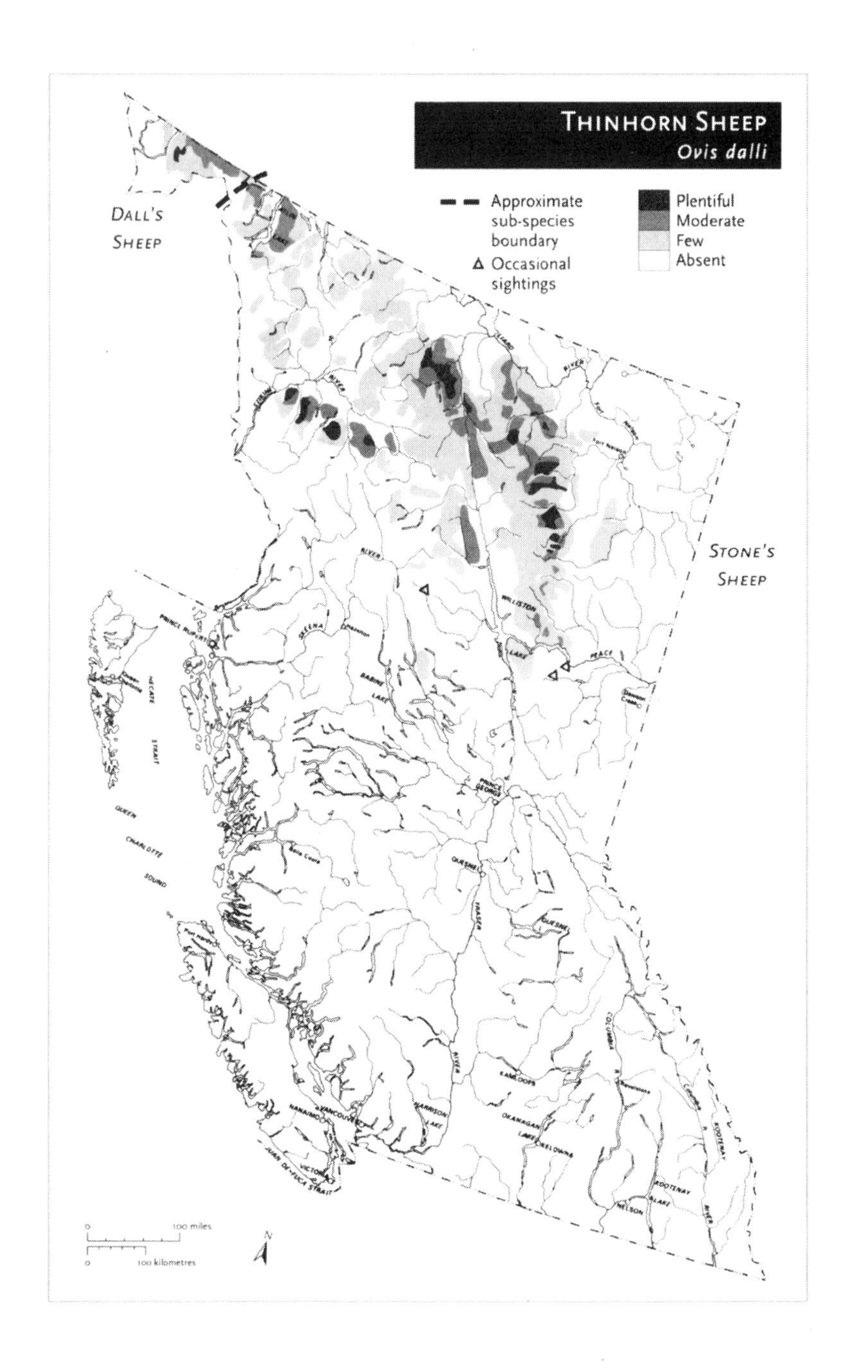
THINHORN SHEEP
Ovis dalli
Approximate sub-species boundary
Occasional sightings
Plentiful
Moderate
Few
Absent
DALL'S SHEEP
STONE'S SHEEP
PRINCE RUPERT
HECATE STRAIT
QUEEN CHARLOTTE SOUND
PRINCE GEORGE
QUESNEL
KAMLOOPS
VANCOUVER
NANAIMO
VICTORIA
JUAN DE FUCA STRAIT
KELOWNA
NELSON
0 100 miles
0 100 kilometres

colour of the neck and face also varies from light grey to darker grey in the dark-colour forms. Stone's Sheep has white inside the ears and grey on the outside, a white belly, white on the backs of the legs and a large white rump patch with a black tail. The dark tail stripe extending forward from the tail may or may not be continuous across the rump patch. Older males may have a dark band running across the underbelly. The horns of both sexes are similar to those of Dall's Sheep.

Measurements:

weight -
  male: 88.6 kg (77-100) n=2
  female: 51.3 kg (45-61) n=7
total length -
  male: 1,595 mm (1,321-1,803) n=5
  female: 1,308 mm (1,295-1,321) n=2
tail vertebrae -
  male: 102 mm (76-114) n=5
  female: 89 mm (76-102) n=2
hind foot length -
  male: 433 mm (419-457) n=5
  female: 380 mm (350-400) n=16
shoulder height -
  male: 1,106 mm n=1
  female: 883 mm (700-950) n=16
ear length -
  male: 95 mm n=1
skull length -
  male: 253.3 mm (250-257) n=5
skull width -
  male: 118.8 mm (117-121) n=5

Range

Stone's Sheep distribution in northern BC runs northwest to southeast from the east side of Bennett Lake on the BC-Yukon border, along the east side of the northern Coast Mountains, into the northern end of the Skeena Mountains, through the Cassiar and Omineca Mountains, and in the northern Rockies to about 80 km northwest of Chetwynd. There is also an unconfirmed report of an isolated population further southwest in the Omineca Range on the east side of Takla Lake. The easternmost population is along the west side of the Chief River, west of the Nelson River. The North American distribution of Stone's Sheep also extends north into an area of integration with Dall's Sheep in south-central Yukon (Cassiar and Pelly mountains, MacArthur Range, and White Mountains).

Between 1990 and 1993, 28 Stone's Sheep were transplanted from the north side of the Peace Arm of the Williston Reservoir to the Upper Moberly River Drainage, but so far the new population has had difficulty becoming established. In another introduction made between 1994 and 1995, 24 Stone's Sheep from the east side of Atlin Lake (Surprise Lake area) were released on the west side of Atlin Lake (Table Mountain). This transplant appears to have been successful and numbers have since increased with as many as 50 animals counted in 1997. The most recent transplant was of 8 sheep from Toad River that were moved to a site 25 km to the east.

Places with reasonably good prospects of observing Stone's Sheep are: the Spatsizi Plateau Wilderness Park; south of Boya Lake on Highway 37 near Good Hope Lake; along the Alaska Highway in the Muncho Lake Park area, where there is a mineral lick; and between Stone Mountain Provincial Park and Muncho Lake near where the Liard River crosses the highway.

Conservation Status

Most of the world's population of Stone's Sheep lives in northern BC, where between 9,500 and 14,400 were estimated in 2011. Until 1997, this subspecies was on the province's Blue List of species at risk, but in 1998, it was downlisted to the Yellow List.

Taxonomy

Populations with intergrades between Dall's and Stone's Sheep occur in northwestern British Columbia and also in the Yukon. In the southern Yukon, I observed individuals in the same group that had all-white coats, white coats and a black tail, and others like the light grey form of Stone's Sheep. Such non-white forms were originally referred to as Fannin's Sheep, but are now simply regarded as intergrades between Dall's and Stone's Sheep. A third subspecies of Thinhorn Sheep recognized by Cowan in 1940 was the Kenai Sheep (*Ovis dalli kenaiensis*) of the Kenai Peninsula, Alaska. Today, this is considered another form of Dall's Sheep. The type locality of Dall's Sheep is the mountains west of the Yukon River, Alaska. For Stone's Sheep, the type locality is the Che-on-nee Mountains at the headwaters of the Stikine River, northwestern BC, but as Dr Ian MacTaggart Cowan pointed out, this name is not marked on any maps, and the locality is more likely the Rainbow Mountains between the Stikine and Iskut rivers. Recent genetic studies support the current division into two subspecies – *dalli* and *stonei*.

Traditional Aboriginal Use

In the northern regions of the province where snowshoes were a main form of winter transportation, First Peoples made the fine babiche netting of snowshoes from the hides of Thinhorn Sheep. They also used the horns to make utensils such as bowls, ladles and spoons, as well as oil holders, knife handles and bracelets. They fashioned bones into drill handles, combs and knives.

Remarks

The possessive "Dall's" and "Stone's" are the correct terms for the subspecies, because both were named after people. Dall's Sheep is named for the American zoologist W.H. Dall, and Stone's Sheep for the Montana naturalist A.J. Stone who discovered this sheep.

Selected References: Bowyer and Leslie 1992, Bunnell and Olsen 1976, Corti and Shackleton 2002, Cowan 1940, Frid 1997, Geist 1971, Hatter and Blower 1996, Nichols 1978, Seip 1983, Worley et al. 2004.

APPENDIX 1
Scientific Names of Plants and Animals Mentioned in this Book

Plants

Lichens

Beard lichens	*Usnea* species
Horsehair lichens	*Bryoria* species
Witch's Hair	*Alectoria sarmentosa*

Grasses and sedges

Common Eel-grass	*Zostera marina*
Reedgrasses	*Calamagrostis* species
Slough Sedge	*Carex obnupta*

Ferns

Deer Fern	*Blechnum spicant*
Sword Fern	*Polystichum munitum*

Forbs and other species

Bur-reeds	*Sparganium* species
Cow-parsnip	*Heracleum maximum*
Pondweeds	*Potamogeton* species
Skunk-cabbage	*Lysichiton americanus*
Swamp Horsetail	*Equisetum fluviatile*

Shrubs and Trees

Amabilis Fir	*Abies amabilis*
Antelope-brush	*Purshia tridentata*

Blackberries	*Rubus* species
Black Cottonwood	*Populus balsamifera trichocarpa*
Black Spruce	*Picea mariana*
Douglas-fir	*Pseudotsuga menziesii*
Douglas Maple	*Acer glabrum*
Dull-leaved Oregon-grape	*Mahonia nervosa*
False Box	*Paxistima myrsinites*
Fireweed	*Epilobium angustifolium*
Hazelnut	*Corylus cornuta*
Highbush-cranberry	*Viburnum edule*
Kinnikinnick	*Arctostaphylos uva-ursi*
Mock-orange	*Philadelphus lewisii*
Mountain-ash	*Sorbus* species
Pacific Ninebark	*Physocarpus capitatus*
Paper Birch	*Betula papyrifera*
Ponderosa Pine	*Pinus ponderosa*
Poplars	*Populus* species
Raspberries	*Rubus* species
Red Elderberry	*Sambucus racemosa*
Red Huckleberry	*Vaccinium parvifolium*
Red-osier Dogwood	*Cornus stolonifera*
Salal	*Gaultheria shallon*
Salmonberry	*Rubus spectabilis*
Saskatoon	*Amelanchier alnifolia*
Soapberry	*Sheperdia canadensis*
Subalpine Fir	*Abies lasiocarpa*
Trembling Aspen	*Populus tremuloides*
Thimbleberry	*Rubus parviflorus*
Western Hemlock	*Tsuga heterophylla*
Western Redcedar	*Thuja plicata*
White Spruce	*Picea glauca*
Willows	*Salix* species

Animals

Invertebrates and disease organisms

American Liver Fluke	*Fascioloides magna*
Anthrax	*Bacillus anthracis*
Biting Gnat	*Culicoides variipennis*
Bot flies	*Cephenemyia* species
Brucellosis	*Brucella abortus*

Lungworms	*Protostrongylus* species
Meningeal (or Brain) Worm	*Parelaphostrongylus tenuis*
Caribou Warble Fly	*Oedemagna tarandi*
Winter Tick	*Dermacentor albipictus*
Tuberculosis	*Mycobactrium bovis*

Vertebrates

Alpine Ibex	*Capra ibex*
American Pronghorn	*Antilocapra americana*
Arabian Camel	*Camelus dromedarius*
Bactrian Camel	*Camelus bactrianus*
Black Bear	*Ursus americanus*
Black-billed Magpie	*Pica hudsonia*
Bobcat	*Lynx rufus*
Chamois	*Rupicapra* species
Chinese Water Deer	*Hydropotes inermis*
Common Eland	*Taurotragus oryx*
Cougar	*Puma concolor*
Coyote	*Canis latrans*
Dik-diks	*Madoqua* species
Domestic Cattle*	*Bos taurus*
Domestic Goat*	*Capra hircus*
Domestic Sheep*	*Ovis aries*
European Bison or Wisent	*Bison bonasus*
European Moose	*Alces alces alces*
European Red Deer	*Cervus elaphus elaphus*
Florida Keys White-tailed Deer	*Odocoileus virginianus clavium*
Four-horned Antelope	*Tetracerus quadricornus*
Giant Bison†	*Bison latifrons*
Giraffe	*Giraffa camelopardalis*
Golden Eagle	*Aquila chrysaetus*
Gorals	*Naemorhedus* species
Greater Kudu	*Tragelaphus strepsiceros*
Grey Wolf	*Canis lupus*
Grizzly Bear	*Ursus arctos*
Harrington's Mountain Goat†	*Oreamnos harringtonii*
Helmeted Muskox†	*Symbos cavifrons*
Indian Blackbuck	*Antilope cervicapra*
Irish Elk†	*Megaloceros (Megaceros) giganteus*

---

* See also Checklist, pages 74-75.

† Extinct species.

Jefferson's Ground Sloth†	*Megalonyx jeffersonii*
Lynx	*Lynx canadensis*
Manitoba Elk	*Cervus elaphus manitobensis*
Mastodon†	*Mammut americanum*
Musk Deer	*Moschus leucogaster*
Muskox	*Ovibos moschatus*
Northern Woodland White-tailed Deer	*Odocoileus virginianus borealis*
Okapi	*Okapia johnstoni*
Persian Fallow Deer	*Dama mesopotamica*
Pronghorn	*Antilocapra americana*
Serow	*Capricornis sumatraensis*
Tule Elk	*Cervus elaphus nannodes*
Wild Boar	*Sus scrofa*
Wolverine	*Gulo gulo*
Woolly Mammoth†	*Mammuthus primigenius*

† Extinct species.

APPENDIX 2
Estimates of Ungulate Numbers in British Columbia in 2011

These 2011 ranges of hoofed mammal populations in BC were compiled by the BC Fish, Wildlife and Habitat Management Branch, Ministry of Forests, from estimates given by regional biologists.

Species	Numbers
Rocky Mountain Elk	38,100–71,900
Roosevelt Elk	5,900–7,200
European Fallow Deer	1,000–1,500
Moose	140,000–235,000
Black-tailed Deer	99,000–155,000
Rocky Mountain Mule Deer	115,000–205,000
White-tailed Deer	87,000–140,000
Woodland Caribou	16,000–27,000
Plains Bison	1,100–1,800
Wood Bison	400–600
Mountain Goat	41,000–66,000
Rocky Mountain Bighorn Sheep	5,900–7,200
Dall's Sheep	400–600
Stone's Sheep	9,500–14,400

APPENDIX 3
Summary of Birth Data for the Hoofed Mammals of British Columbia

Species	Gestation (days)	Birth Weight (kg)	Birth Period	Young's Strategy*
Elk	255	10–15	late May – early June	hider
Fallow Deer	150–220	2–4	late May – June	hider
Moose	240–246	11–16	June	?
Mule Deer	203	2–4	June	hider
White-tailed Deer	200–210	2–4	early June	hider
Woodland Caribou	228	5–12	late May – early June	follower
Bison	277–293	14–18	late April – June	follower
Mountain Goat	147–178	2–3	late May – early June	follower
Bighorn Sheep	175	3–6	May – June	follower
Thinhorn sheep	175	3–4	May – June	follower

* See page 49.

GLOSSARY

Amino acids Constituents of proteins.

Ano-genital The region around the anus and genital area.

Artiodactyl A hoofed mammal that has an even number of toes.

Biodiversity Biological diversity: the variety of living organisms in an area, including both genetic and species diversity.

Biogeoclimatic zone An area with a relatively homogeneous climate and characteristic vegetation; British Columbia has 14 biogeoclimatic zones. See http://www.for.gov.bc.ca/hfd/library/documents/treebook/biogeo/biogeo.htm

Brachydont (also **Brachyodont**) Cheek teeth in which the crown is about equal in height to the roots (i.e., low-crowned), usually associated with browsers and omnivores in ungulates.

Broomed Broken and frayed, as the horn tips of Bighorn Sheep become after engaging in many fights with rivals.

Browser A herbivore that eats mainly the leaves and other parts of woody plants (i.e., browse).

Bunodont Cheek teeth with crowns formed by low conical cusps covered by enamel, typical of cheek teeth of non-herbivorous mammals.

Caecum The blind pouch at the junction of the small and large intestines, often the site of microbial fermentation in herbivores.

Canines The teeth between the incisors and premolars. Most mammals have a single canine on each side of the lower and upper jaws, but some species of herbivorous ungulates lack the upper pair, and their lower canines are incisiform.

Carpals Bones of the forefoot at the distal end of the radius and ulna that comprise the wrist.

Cementum annuli Cementum in the roots of the teeth is laid down seasonally, probably in response to changes in nutrition. This differential growth creates distinct rings that can be used to determine the age of an animal.

Cheek teeth The teeth located behind the canines: premolars and molars.

Conspecific Individuals or populations belonging to the same species.

Cover Vegetation or topographic features used by animals either for protection against the weather (thermal cover) or against predators (security cover).

Cursorial A type of locomotion adapted for running fast.

Deciduous dentition The teeth (incisors, canines and premolars) of juvenile animals that are replaced by permanent teeth.

Dentition The form and arrangement of the teeth.

Dew claws The lateral toes and hooves that are much smaller in size and located higher than the main hooves. In most species they bear little if any of the animal's weight.

Diastema The space on the lower jaw of ungulates between the lower canine and the first of the premolars.

Digestion The breakdown of food by enzymes into simpler chemical components that can be absorbed in the animal's digestive system.

Digits The terminal bones of the limbs forming the fingers and toes.

Distal Farthest from the body, usually referring to parts of bones or limbs (opposite to proximal).

Ecoprovince A broad geographic area with consistent climate and terrain; British Columbia has nine terrestrial ecoprovinces (see the map on page 52).

Ecotype A subdivision (e.g., a population or group of populations) within a species or subspecies that is considered to be adapted to a particular set of local environmental conditions. The term has been applied mainly to plants.

Enzymes Chemicals produced in the digestive system that break down food so that it can be absorbed.

Epidermis The outer layer of the skin.

Escape terrain The security cover used by Mountain Goats and mountain sheep, typically comprised of steep bluffs and rock cliffs.

Esophagus The first part of the alimentary canal or digestive system, between the pharynx and the stomach or fore-stomach. In Artiodactyls, the lower part of the esophagus is enlarged into one or more chambers in front of the true stomach.

Estrus A period of sexual receptivity in female mammals; heat.

Eutherian A mammal that nourishes its unborn young through a placenta. (Eutheria is an infraclass in the Class Mammalia.)

Fecal pellet The hard, lozenge-shaped feces of many ungulates. The fresh feces of most BC ungulates, except for Bison, Domestic Cattle and Horses, are fecal pellets.

Feces The remains of food excreted by an animal after it has passed through its digestive tract. Feces also includes cells from the gut lining. Feces are also called droppings, scats and pellets.

Feral Former domestic species now living in the wild, free of human husbandry.

Fermentation In herbivores, the anaerobic digestion of food in the digestive system by micro-organisms such as protozoa and bacteria.

Foramen (plural: **Foramina**) A small opening or perforation in a bone.

Forbs Non-woody, broad-leaved flowering plants (Dicotyledons).

Founder effect A population with low genetic diversity because it was initiated (founded) by only a few animals that either moved or were introduced into the area.

Gestation period The length of pregnancy; the time from fertilization of the ovum to the birth of the fetus.

Glands Epithelial tissues that produce volatile, odoriferous chemicals. An ungulate uses glandular secretions primarily to communicate with others of its own species.

Grazer A herbivorous mammal that eats mainly grasses, sedges and forbs.

Haida Gwaii Islands off the coast of central British Columbia, formerly known as the Queen Charlotte Islands.

Harem A group of animals in the mating season composed of a single adult male and two or more adult females. Often immature individuals are present, but no other sexually active males. The adult male defends the females against other males and courts the females in the group.

Home range The geographic area used by an animal during its normal day-to-day activities; usually measured over a year.

Hypsodont High-crowned cheek teeth, usually associated with grazing herbivores.

Incisiform Formed like incisors, referring to the lower canine teeth of herbivorous ungulates that are shaped like incisors and serve to increase the size of the cropping surface used for gathering food.

Incisors The front teeth of the mammalian upper and lower jaws. There are usually three on each side of the upper and lower jaws. Many Artiodactyls lack upper incisors.

Keratin A fibrous protein that forms the basis of epidermal structures such as hairs, horns and nails.

Keratinize The process by which the epidermis changes to keratin.

Lophodont An enamel pattern of cheek teeth where the enamel has been folded (mainly laterally) and when the top surface wears away, it reveals a convoluted pattern of vertically oriented ridges. Many herbivorous mammals, such as horses, rodents and rabbits have teeth with lophodont enamel.

Metabolize To bring about chemical change, either by creating or breaking down compounds, in the cells of living organisms.

Metapodials The bones lying between the digits and the carpals or tarsals. Those of the front limbs are called metacarpals and those of the hind limb metatarsals.

Molars The rear cheek teeth located behind the premolars in the mammalian jaw. Ungulates have three molars on each side of the upper and lower jaws. The molars are never shed.

Morphology The form and physical characteristics of an organism or its parts.

Nares The openings of the nasal passages in the skull, both front (anterior) and rear (posterior); the nostrils.

Neonate Newborn young, usually no more than one day old.

Parasite An organism that obtains nutrients from another organism (the host) within which it lives, usually without causing the host's death.

Pedicel The short bone extension on the frontal bone of the skull from which the antlers grow; also called a pedicle.

Pelage The fur or hair covering of a mammal.

Perissodactyl A hoofed mammal that has an odd number of toes.

Phalanges The individual bones of the fingers and toes.

Plantigrade A form of locomotion in which the entire length of the metapodials, along with the digits, are in contact with the ground; typical of bears and humans.

Pneumation An expansion of the skull sinuses (e.g., the frontal and cornual sinuses) creating spaces between the bone covering the brain and the outer skull bones, with thin bone connections between these two plates. Pneumation is found in several species that fight by butting or clashing head-to-head; it likely helps absorb the impact.

Polygamous A mating system in which both sexes mate with more than one partner in each mating season.

Polygynous A mating system in which the male mates with more than one female but the female usually mates with only one male in each mating season.

Precocial Newborn animals that are relatively well developed at birth. Their eyes are open and usually they can walk and move about effectively shortly after birth.

Premolar The cheek teeth located between the canines and the molars in the mammalian jaw. Most ungulates have three premolars on each side of the upper and lower jaws.

Primitive A species or character that is considered to be an early or ancestral form; it also applies to a characteristic of a species that is more typical of ancestral forms.

Proximal Closest to the body, usually referring to parts of the bones or limbs (opposite to distal).

Riparian Of or on the area around a stream or river.

Ruminate To regurgitate food from the fore-stomach (rumen-reticulum) to the mouth where it is re-chewed, mixed with saliva and re-swallowed.

Rut The mating period of ungulates when females come into estrus and conception takes place.

Selenodont A cheek tooth enamel pattern in which the surface enamel has been folded (mainly vertically) and the top surface worn away to leave an elliptical or crescent-shaped loop of vertical oriented ridges. Herbivorous Artiodactyls have selenodont teeth.

Subspecies Populations of species that are geographically separated and differ taxonomically from other populations; sometimes called geographic races.

Subunguis The softer tissue immediately behind the keratinized hoof (unguis), continuous with the pad at the end of the digit. The subunguis provides traction for locomotion.

Symbiotic relationship A relationship between two species in which both benefit.

Tarsals Bones of the hind foot at the distal end of the tibia and fibula that comprise the ankle.

Tending pair A single adult male guarding and courting a single adult female.

Tine A branch from the main beam of an antler. Sometimes also called an antler point or spike.

Unguis The outer hard, keratinized epidermal sheath of the hoof that is in direct contact with the ground. It helps protect the underlying subunguis.

Unguligrade A form of locomotion in which only the tips of the hooves rest on the ground, typical of ungulates.

Vacuity An opening or hollow.

Vertebrate Animals with spinal columns or backbones (either of bone or cartilage).

Vestigial A part of the body that has degenerated or atrophied in the course of evolution.

Wallow A depression in the ground filled with dust, mud or water where animals will roll about; to roll about in a wallow. Many animals make their own wallows by pawing, digging and rolling in the same place every year.

REFERENCES

Anderson, A.E., and O.C. Wallmo. 1984. *Odocoileus hemionus. Mammalian Species* 219.

Armleder, H.M., M.J. Waterhouse, D.G. Keisker and R.J. Dawson. 1994. Winter habitat use by Mule Deer in the central interior of British Columbia. *Canadian Journal of Zoology* 72:1721–25.

Banfield, A.W.F. 1961. A revision of the Reindeer and Caribou, Genus *Rangifer*. National Museum of Canada, *Bulletin* No. 177, Biological Series No. 66.

Berger, J. 1986. *Feral Horse Ecology. A Review of Wild Horses of the Great Basin: Social Competition and Population Size*. Chicago: University of Chicago Press.

Berger, J., and C. Cunningham. 1994. *Bison: Mating and Conservation in Small Populations*. New York: Columbia University Press.

Bergerud, A.T. 1996. Evolving perspectives on Caribou population dynamics: have we got it right yet? *Rangifer* Special Issue 9:95–115.

Bleich, V.C., R.T. Bowyer and J.D. Wehausen. 1997. Sexual segregation in mountain sheep: resources or predation? *Wildlife Monographs* 134.

Blood, D.A., J.R. McGillis and A.L. Lovaas. 1967. Weights and measurements of Moose in Elk Island National Park, Alberta. *Canadian Field-Naturalist* 81:263–69.

Blood, D.A., and G.W. Smith. 1984. Weights and measurements of Roosevelt Elk on Vancouver Island. *The Murrelet* 65:41–44.

Bowyer, R.T., and D.M. Leslie, Jr. 1992. *Ovis dalli. Mammalian Species* 393.

Brunt, K. 1990. Ecology of Roosevelt Elk. In: *Deer and Elk Habitats in Coastal Forests of Southern British Columbia*, edited by J.B. Nyberg and D.W. Janz. Victoria: BC Ministry of Forests and BC Ministry of Environment.

Bubenik, G.A., and A.B. Bubenik, editors. 1990. *Horns, Pronghorns and Antlers: Evolution, Morphology, Physiology and Social Significance*. New York: Springer Verlag.

Bunnell, F.L. 1990. Ecology of Black-tailed Deer. In *Deer and Elk Habitats in Coastal Forests of Southern British Columbia*, edited by J.B. Nyberg and D.W. Janz. Victoria: BC Ministry of Forests and Ministry of Environment.

Bunnell, F.L., and N.A. Olsen. 1976. Weights and growth of Dall Sheep in Kluane Park Reserve, Yukon Territory. *Canadian Field-Naturalist* 90:157–62.

Cannings, R., and S. Cannings. 1996. *British Columbia: A Natural History*. Vancouver: Greystone Books.

Carl, G.C., and C.J. Guiguet. 1972. *Alien Animals in British Columbia*, 2nd ed. Handbook 14. Victoria: British Columbia Provincial Museum.

Caro, T.M., C.M. Graham, C.J. Stoner, and J.K. Vargas. 2004. Adaptive significance of antipredator behaviour in artiodactyls. *Animal Behaviour* 67:205–28.

Carr, S.M., and G.A. Hughes. 1993. The direction of introgressive hybridization between species of North American deer (*Odocoileus*) as inferred from mitochondrial cytochrome B sequences. *Journal of Mammalogy* 74:331–32.

Chabot, D. 1993. Communication. In *Hoofed Mammals of Alberta*, edited by J.B. Stelfox. Edmonton: Lone Pine Publishing.

Chapman, D., and N. Chapman. 1975. *Fallow Deer: Their History, Distribution and Biology*. Lavenham, Suffolk: Terrence Dalton.

Cichowski, D.B. 1993. Seasonal movements, habitat use and winter feeding ecology of Woodland Caribou in west-central British Columbia. BC Ministry of Forests, *Land Management Report* 79.

Coltman, D.W., P. O'Donoghue, J.T. Jorgenson, J.T. Hogg, C. Strobeck and M. Festa-Bianchet. 2003. Undesirable evolutionary consequences of trophy hunting. *Nature* 426: 655–58.

Corti, P., and D.M. Shackleton. 2002. Relationship between predation risk factors and sexual segregation in Dall's Sheep (*Ovis dalli dalli*). *Canadian Journal of Zoology* 80:2108–17.

Côté, S.D. 1996. Mountain Goat responses to helicopter disturbance. *Wildlife Society Bulletin* 24:681–85.

Côté, S. D., M. Festa-Bianchet and F. Fournier. 1998. Life-history effects of chemical immobilization and radio collars in Mountain Goats. *Journal of Wildlife Management* 62:745–52.

Cowan, I. McTaggart. 1936. Distribution and variation in deer (Genus *Odocoileus*) of the Pacific coastal region of North America. *California Fish & Game* 22:155–246.

Cowan, I. McTaggart. 1940. Distribution and variation in the native sheep of North America. *American Midland Naturalist* 24:505–80.

Cowan, I. McTaggart. 1951. The diseases and parasites of big game mammals of western Canada. *Proceedings of the Annual Game Convention* 5:37–64

Cowan, I. McTaggart. 1987. Science and the conservation of wildlife in British Columbia. In *Our Wildlife Heritage: 100 Years of Wildlife Management*, edited by A. Murray. Victoria: The Centennial Wildlife Society of British Columbia.

Cowan, I. McTaggart, and C.J. Guiguet. 1965. *The Mammals of British Columbia*, rev. ed. Handbook 11. Victoria: British Columbia Provincial Museum.

Cowan, I. McTaggart, and W. McCrory. 1970. Variation in the Mountain Goat, *Oreamnos americanus* (Blainville). *Journal of Mammalogy* 51:60–73.

Cronin, M.A. 1991. Mitochondrial and nuclear genetic relationships of deer (*Odocoileus* spp.) in western North America. *Canadian Journal of Zoology* 69:1270–79.

Cronin, M.A. 1992. Intraspecific mitochondrial DNA variation in North American cervids. *Journal of Mammalogy* 73:70–82.

Darwin, C. 1899. *The Descent of Man and Selection in Relation to Sex*, 2nd ed. London: John Murray.

Demarais, S., and P.R. Krausman. 2000. *Ecology and Managament of Large Mammals in North America.* Upper Saddle River, New Jersey: Prentice-Hall.

Demarchi, D.A., and R.A. Demarchi. 1987. Wildlife habitat – the impacts of settlement. In *Our Wildlife Heritage: 100 Years of Wildlife Management*, edited by A. Murray. Victoria: The Centennial Wildlife Society of British Columbia.

Duffy, M.S., T.A. Greaves, N.J. Keppie and M.D.B. Burt. 2002. Meningeal worm is a long-lived parasitic nematode in White-tailed Deer. *Journal of Wildlife Diseases* 38:448–52.

Eccles, T.R., and D.M. Shackleton. 1979. Recent records of twinning in mountain sheep. *Journal of Wildlife Management* 43:974–76.

Feldhamer, G.A., K.C. Farris-Renner and C.M. Barker. 1988. *Dama dama. Mammalian Species* 317.

Feldhamer, G.A., L.C. Drickamer, S.H. Vessey and J.F. Merritt. 1999. *Mammalogy: Adaptation, Diversity and Ecology*. New York: WCB/McGraw-Hill.
Festa-Bianchet, M., and S.D. Côté. 2008. *Mountain Goats: Ecology, Behavior and Conservation of an Alpine Ungulate*. Washington, DC: Island Press.
Franzmann, A.W., and C.C. Schwartz, editors. 1998. *Ecology and Management of the North American Moose*. A Wildlife Management Institute Book. Washington, DC: Smithsonian Institution.
Frid, A. 1997. Vigilance by female Dall's Sheep: interactions between predation risk factors. *Animal Behaviour* 53:799–808.
Geist, V. 1963. On the behaviour of the North American Moose (*Alces alces andersoni* Peterson 1950) in British Columbia. *Behaviour* 20:377–416.
Geist, V. 1964. On the rutting behaviour of the Mountain Goat. *Journal of Mammalogy* 45:551–68.
Geist, V. 1966. The evolution of horn-like organs. *Behaviour* 27:177–214.
Geist, V. 1971. *Mountain Sheep: A Study in Behavior and Ecology*. Chicago: University of Chicago Press.
Geist, V. 1986. New evidence of high frequency antler wounding in cervids. *Canadian Journal of Zoology* 64, 380-384.
Geist, V. 1990. *Mule Deer Country*. Saskatoon: Western Producer Prairie Books.
Geist, V. 1991. Phantom subspecies: The Wood Bison, *Bison bison* "*athabascae*" Rhoads 1897, is not a taxon, but an ecotype. *Arctic* 44:283–300.
Geist, V., and P. Karsten. 1977. The Wood Bison (*Bison bison athabascae* Rhoads) in relation to the hypothesis on the origin of the American Bison (*Bison bison bison*). *Zeitschrift für Säugetierkunde* 42:119–27.
Goldman, E.A., and R. Kellog. 1940. Ten new White-tailed Deer from North and Middle America. *Proceedings of the Biological Society of Washington* 53:81–90.
Gosling, L.M. 1985. The even-toed ungulates: order Artiodactyla: sources, behavioural context, and function of chemical signals. In *Social Odours in Mammals*, vol. 2, edited by R.T.E. Brown and D.W. Macdonald. Oxford, UK: Clarendon Press.
Goss, R.J. 1983. *Deer Antlers: Regeneration, Function and Evolution*. New York: Academic Press.
Guthrie, R.D. 1990. *Frozen Fauna of the Mammoth Steppe: The Story*

of Blue Babe. Chicago: University of Chicago Press.
Halls, L.K., editor. 1984. *White-tailed Deer: Ecology and Management*. The Wildlife Institute. Harrisburg, PA: Stackpole Books.
Harington, C.R. 1996. Quaternary animals: vertebrates of the Ice Age. In *Life in Stone: A Natural History of British Columbia's Fossils*, edited by R. Ludvigsen. Vancouver: UBC Press.
Harper, B., S. Cannings, D. Fraser and W.T. Munro. 1994. Provincial lists of species at risk. In *Biodiversity in British Columbia*, edited by L.E. Harding and E. McCullum. Delta, BC: Canadian Wildlife Service.
Hatter, I.W., and D. Blower. 1996. History of transplanting Mountain Goats and Mountain Sheep – British Columbia. *Proceedings of the Biennial Symposium of the Northern Wild Sheep & Goat Council* 10:158–63.
Heard, D.C., and K.L. Vagt. 1998. Caribou in British Columbia: a 1996 status report. *Rangifer* Special Issue 10:117–23.
Hebert, D., and S. Harrison. 1988. The impact of Coyote predation on lamb mortality patterns at the Junction Wildlife Management Area. *Proceedings of the Biennial Symposium of the Northern Wild Sheep & Goat Council* 5:283–91.
Hebert, D.M., and T. Smith. 1986. Mountain Goat management in British Columbia. *Proceedings of the Biennial Symposium of the Northern Wild Sheep & Goat Council* 5:48–59.
Hesselton, W.T., and R.M. Hesselton. 1983. White-tailed Deer *Odocoileus virginianus*. In *Wild Mammals of North America*, edited by J.A. Chapman and G.A. Feldhamer. Baltimore, MD: The Johns Hopkins University Press.
Holm, B., and B. Reid. 1975. *Indian Art of the Northwest Coast: A Dialogue on Craftmanship and Aesthetics*. Houston: Institute for the Arts, Rice University.
Hudson, R.J., and R.G. White. 1985. *Bioenergetics of Wild Herbivores*. Boca Raton, Florida: CRC Press.
Johnson, D.R., and D.W. Nagorsen. 1990. Evaluation of cranial and antler characteristics to determine sex of Mountain Caribou. *Canadian Field-Naturalist*. 104:583–84.
Jones, R.L., and H.C. Hanson. 1985. *Mineral Licks, Geophagy and Biogeochemistry of North American Ungulates*. Ames: Iowa University Press.
Klaus, G., and B. Schmidg. 1998. Geophagy at natural licks and mammal ecology: a review. *Mammalia* 62:482–89.
Krebs, J. R., and N.B. Davies. 1993. *An Introduction to Behavioural Ecology*, 3rd ed. Oxford, UK: Blackwell Scientific Publications.

Kunkel, K.E., and D.H. Pletscher. 2000. Habitat factors affecting vulnerability of Moose to predation by wolves in southeastern British Columbia. *Canadian Journal of Zoology* 78:150–52.

Lankester, M.W., and W.M. Samuel. 1998. Pests, parasites and diseases. In *Ecology and Management of the North American Moose*, edited by A.W. Franzmann and C.C. Schwartz. A Wildlife Management Institute Book. Washington, DC: Smithsonian Institution.

Larter, N.C., and C.C. Gates. 1990. Home ranges of Wood Bison in an expanding population. *Journal of Mammalogy* 71:604–607.

Lawrence, P.K., S. Shanthalingam, R.P. Dassanayake, R. Subramaniam, C.N. Herndon, D.P. Knowles, F.R. Rurangirwa, W.J. Foreyt, G. Wayman, A-M. Marciel, S.K. Highlander and S. Srikumaran. 2010. Transmission of *Mannheimia haemolytica* from Domestic Sheep (*Ovis aries*) to Bighorn Sheep (*Ovis canadensis*): unequivocal demonstration with green fluorescent protein-tagged organisms. *Journal of Wildlife Diseases* 46:706–17.

Lent, P.C. 1974. A review of the rutting behaviour in Moose. *Le Naturaliste Canadien* 101:307–323.

Leopold, A. 1942. *Game Management*, 2nd ed. New York: C. Scribner's Sons.

Leslie, D.M., and K.J. Jenkins. 1985. Rutting mortality among male Roosevelt Elk. *Journal of Mammalogy* 66:163–64.

Lott, D.F. 1974. Sexual and aggressive behaviour of adult male American Bison (*Bison bison*). In *The Behaviour of Ungulates and its Relation to Management*, edited by V. Geist and F. Walther. IUCN Publications New Series No. 24. Morges, Switzerland: International Union for the Conservation of Nature and Natural Resources.

Macdonald, D., editor. 1984. *The Encyclopaedia of Mammals*. New York: Facts On File Publications; and Oxford, UK: Equinox.

Mackie, R.J., K.L. Hamilin and D.F. Pac. 1982. Mule Deer, *Odocoileus hemionus*. In *Wild Mammals of North America*, edited by J.A. Chapman and G.A. Feldhamer. Baltimore, MD: The Johns Hopkins University Press.

Main, M.B., F.W. Weckerly and V.C. Bleich. 1996. Sexual segregation in ungulates: new directions for research. *Journal of Mammalogy* 77:449–61.

McCullough, D.R. 1965. Sex characteristics of Black-tailed Deer hooves. *Journal of Wildlife Management* 29:210–12.

McNay, R.S., and J.M. Voller. 1995. Mortality causes and survival estimates for adult female Columbian Black-tailed Deer. *Journal of Wildlife Management* 59:138–46.

Meagher, M. 1986. *Bison bison*. *Mammalian Species* 266.

Miller, F.L. 1982. Caribou, *Rangifer tarandus*. In *Wild Mammals of North America*, edited by J.A. Chapman and G.A. Feldhamer. Baltimore, MD: The Johns Hopkins University Press.

Moody, A., B. Burtin and R. Moody. 1994. *An Investigation into the Fallow Deer of Sidney Spit Provincial Park*. Occasional Paper No. 1. Victoria: Parks Branch, BC Ministry of Environment.

Moore, W.G., and R.L. Marchinton. 1974. Marking behaviour and its social function in White-tailed Deer. In *The Behaviour of Ungulates and its Relation to Management*, edited by V. Geist and F. Walther. IUCN Publications New Series No. 24. Morges, Switzerland: International Union for the Conservation of Nature and Natural Resources.

Müller-Schwartz, D. 1991. The chemical ecology of ungulates. *Applied Animal Behaviour Science* 29:389–402.

Murie, O.J. 1975. *A Field Guide to Animal Tracks*, 2nd ed. The Peterson Field Guide Series. Boston, MA: Houghton and Mifflin.

Nagorsen, D.W. 1990. *The Mammals of British Columbia: A Taxonomic Catalogue*. Memoir No. 4. Victoria: Royal BC Museum.

Nagorsen, D.W. 1996. *Opossums, Shrews and Moles of British Columbia*. Royal BC Museum Handbook. Vancouver: UBC Press.

Nagorsen, D.W., and R.M. Brigham. 1993. *Bats of British Columbia*. Royal BC Museum Handbook. Vancouver: UBC Press.

Naughton, D. 2012. *The Natural History of Canadian Mammals*. Toronto: University of Toronto Press.

Nelson, E.W. 1914. Description of a new subspecies of Moose from Wyoming. *Proceedings of the Biological Society of Washington* 27:71–74.

Nichols, L. 1978. Dall Sheep reproduction. *Journal of Wildlife Management* 42: 570–80.

Nowak, R.M. 2003. *Walker's Mammals of the World*, 6th ed., vol. 2. Baltimore, MD: The Johns Hopkins University Press.

Nyberg, J.B., and D.W. Janz, editors. 1990. *Deer and Elk Habitats in Coastal Forests of Southern British Columbia*. Victoria: BC Ministry of Forests and Ministry of Environment.

Parker, S.P., editor. 1990. *Grzimek's Encyclopedia of Mammals*, vols 4 and 5. New York: McGraw-Hill Publishing.

Peek, J.M. 1974. A review of Moose food habits in North America. *Le Naturaliste Canadien* 101:195–215.

Peterson, R.L. 1952. A review of the living representatives of the genus Alces. *Contributions of the Royal Ontario Museum of Zoology and Paleontology* 34.

Pielou, E.C. 1991. *After the Ice Age: The Return of Life to Glaciated North America*. Chicago: The University of Chicago Press.

Poole, K.G., K. Stuart-Smith and I.E. Teske. 2009. Wintering strategies by Mountain Goats in interior mountains. *Canadian Journal of Zoology* 87: 273–83.

Portier, C., M. Festa-Bianchet, J-M. Gaillard, J.T. Jorgenson and N.G. Yoccoz. 1998. Effects of density and weather on survival of Bighorn Sheep lambs (*Ovis canadensis*). *Journal of Zoology* (London) 245:271–78.

Price, S.A, O.R.P Bininda-Emonds and J.L. Gittleman. 2005. A complete phylogeny of the whales, dolphins and even-toed hoofed mammals (Cetartiodactyla). *Biological Reviews* 80:445–73.

Quayle, J.F., and K.R. Brunt. 2003. *Status of Roosevelt Elk (Cervus elaphus roosevelti) in British Columbia*. Victoria: BC Ministry of Sustainable Resource Management, Conservation Data Centre, and BC Ministry of Water, Land and Air Protection, Biodiversity Branch.

Reynolds, H.W., R.D. Glaholt and A.W.L. Hawley. 1982. Bison (*Bison bison*). In *Wild Mammals of North America*, edited by J.A. Chapman and G.A. Feldhamer. Baltimore, MD: The Johns Hopkins University Press.

Reynolds, H.W., R.M. Hansen and D.G. Peden. 1978. Diets of the Slave River lowlands Bison herd, Northwest Territories, Canada. *Journal of Wildlife Management* 42:581–90.

Rezendes, Paul. 1992. *Tracks and the Art of Seeing: How to Read Animal Tracks and Sign*. Vermont: Camden House Publishing.

Rideout, C.B., and R.S. Hoffmann. 1975. *Oreamnos americanus*. *Mammalian Species* 63.

Robbins, C.T. 1993. *Wildlife Feeding and Nutrition*. New York: Academic Press.

Robinson, D.J. 1987. Wildlife and the law. In *Our Wildlife Heritage: 100 years of Wildlife Management*, edited by A. Murray. Victoria: The Centennial Wildlife Society of British Columbia.

Robinson, H.S., R.B. Wielgus and J.C. Gwilliam. 2002. Cougar predation and population growth of sympatric Mule Deer and White-tailed Deer. *Canadian Journal of Zoology* 80:556–68.

Roe, F.G. 1951. *The North American Buffalo*. Toronto: University of Toronto Press.

Rominger, E.M., and J.L Oldemeyer. 1990. Early winter diet of Woodland Caribou in relation to snow accumulation, Selkirk Mountains, British Columbia, Canada. *Canadian Journal of Zoology* 68:2691–94.

Samuel, W.M. 1993. Parasites and diseases. In *Hoofed Mammals of Alberta*, edited by J.B. Stelfox. Edmonton: Lone Pine Publishing.

Seip, D.R. 1983. Foraging ecology and nutrition of Stone's Sheep. B.C. Ministry of Environment, *Fish and Wildlife Report* No. R-9.

Seip, D.R., and D.B. Cichowski. 1994. Population ecology of Caribou in British Columbia. *Rangifer* Special Issue 9:73–80.

Schonewald-Cox, C.M., J.W. Bayless and J. Schonewald. 1985. Cranial morphometry of Pacific coast Elk (*Cervus elaphus*). *Journal of Mammalogy* 66:63–74.

Schwantje, H.M. 1988. Causes of Bighorn Sheep mortality and die-offs. BC Ministry of Environment, *Wildlife Working Report* 35.

Shackleton, D.M. 1985. *Ovis canadensis*. *Mammalian Species* 230.

Shackleton, D.M., L.V. Hills and D.A. Hutton. 1975. Aspects of variation in cranial characters of Plains Bison (*Bison bison bison* Linneaus) from Elk Island National Park, Alberta. *Journal of Mammalogy* 56:871–877.

Shackleton, D.M., and D.A. Hutton. 1971. An analysis of the mechanisms of brooming in mountain sheep horns. *Zeitschrift für Säugetierkunde* 36:342–50.

Shackleton, D.M., C.C. Shank and B.M. Wikeem. 1999. Natural history of Rocky Mountain and California Bighorn Sheep. In *Mountain Sheep of North America*, edited by R. Valdez and P.R. Krausman. Tucson: University of Arizona Press.

Shank, C.C. 1972. Some aspects of the social behaviour in a population of feral goats (*Capra hircus* L.). *Zeitschrift für Tierpsychologie* 30:488–528.

Shannon, N.H., R.J. Hudson, V.C. Brink and W.D. Kitts. 1975. Determinants of spatial distribution of Rocky Mountain Bighorn Sheep. *Journal of Wildlife Management* 39:387–401.

Silvy, N.J. 2012. *The Wildlife Techniques Manual*, vol. 1: *Research*, vol. 2: *Management*, 7th ed. Bethesda, MD: The Wildlife Society.

Smith, W.P. 1991. *Odocoileus virginianus*. *Mammalian Species* 388.

Spalding, D.J. 1990. The early history of Moose (*Alces alces*): distribution and relative abundance in British Columbia. Royal BC Museum *Contributions to Natural Science* 11.

Spalding, D.J. 1992. The history of Elk (*Cervus elaphus*) in British Columbia. Royal BC Museum *Contributions to Natural Science* 18.

Stelfox, J.B., editor. 1993. *Hoofed Mammals of Alberta*. Edmonton: Lone Pine Publishing.
Stevenson, S.K., K.N. Child, G.S. Watts and E.L. Terry. 1991. The Mountain Caribou in managed forests program: integrating forestry and habitat management in British Columbia. *Rangifer* Special Issue 7:130–36.
Strobeck, C., and J. Coffin. 1996. Genetic relationships between Woodland and Barren Ground Caribou. *Rangifer* Special Issue 9:397.
Sutcliffe, A.J. 1985. *On the Track of Ice Age Mammals*. Cambridge, MA: Harvard University Press.
Telfer, E.S., and J.P. Kelsall. 1984. Adaptations of some large North American mammals for survival in snow. *Ecology* 65:1828–34.
Tinbergen, N. 1953. *Social Behaviour in Animals*. London: Methuen.
van Zyll de Jong, C.G. 1986. A systematic study of recent Bison, with particular consideration to the Wood Bison (*Bison bison athabascae* Rhoads 1898). National Museums of Canada, Ottawa *Publication of Natural Sciences* 6.
Vaughan, T.V., J.M. Ryan and N.J. Czaplewski. 2011. *Mammalogy*, 5th ed. Mississauga, Ont.: Jones and Bartlett Publishers.
Vore, J.M., and E.M. Schmidt. 2001. Movements of female Elk during calving season in northwest Montana. *Wildlife Society Bulletin* 29:720–25.
Walker, A.B.D., K.L. Parker and M.P. Gillingham. 2006. Behaviour, habitat associations and intrasexual differences of female Stone's Sheep. *Canadian Journal of Zoology* 84:1187–1200.
Wallmo, O.C. 1981. *Mule and Black-tailed Deer of North America*. The Wildlife Management Institute. Lincoln: University of Nebraska Press.
Walther, F.R. 1984. *Communication and Expression in Hoofed Mammals*. Bloomington: Indiana University Press.
Ward, P.D. 1997. *The Call of Distant Mammoths: Why the Ice Age Mammals Disappeared*. New York: Copernicus.
Wareham, W. 1991. *British Columbia Wildlife Viewing Guide*. Edmonton: Lone Pine Publishing.
Waterhouse, M., H. Armleder and R. Dawson. 1993. *Winter Food Habits of Mule Deer in the Central Interior of British Columbia*. Technical Report. Victoria: BC Ministry of Forests.
Weyhausen, J.D., and R.R. Ramey II. 2000. Cranial morphometric and evolutionary relationships in the northern range of *Ovis canadensis*. *Journal of Mammalogy* 81:145–61.

Wikeem, B.M., and M.D. Pitt. 1992. Diet of California Bighorn Sheep: assessing optimal foraging habitat. *Canadian Field-Naturalist* 106:327–35.

Wilson, D.E., and D.M. Reeder, editors. 1993. *Mammal Species of The World: A Taxonomic and Geographic Reference*, 2nd ed. Washington, DC: Smithsonian Institution.

Wisdom, M.J., and J.G. Cook. 2000. Elk. In *Ecology and Management of Large Mammals in North America*, edited by S. Demarais and P.R. Krausman. Upper Saddle River, NJ: Prentice-Hall.

Wishart, W.D. 1984. Western Canada. In *White-tailed Deer: Ecology and Management*, edited by L.K. Halls. The Wildlife Institute. Harrisburg, PA: Stackpole Books.

Wittmer, H.U., B.N. McLellan, D.R. Seip, J.A. Young, T.A. Kinley, G.S. Watts and D. Hamilton. 2005a. Population dynamics of the endangered mountain ecotype of Woodland Caribou (*Rangifer tarandus caribou*) in British Columbia. *Canadian Journal of Zoology* 83:407–18.

Wittmer H.U., A.R.E. Sinclair and B.N. McLellan. 2005b. The role of predation in the decline and extirpation of Woodland Caribou. *Oecologia* 144:257–67.

Wood, M.D. 1996. Seasonal habitat use and movements of Woodland Caribou in the Omineca Mountains, north central British Columbia, 1991–1993. *Rangifer* Special Issue 9:365–78.

Worley, K., C.Strobeck, S. Arthur, J. Carey, H. Schwantje, A. Veitch and D.W. Coltman. 2004. Population genetic structure of North American Thinhorn Sheep (*Ovis dalli*). *Molecular Ecology* 13:2545–56.

For more publications on BC's hoofed mammals, along with related topics, see the Biodiversity Publications Catalogue, accessible online at: http://www.env.gov.bc.ca/wld/catalogue/index.html

ACKNOWLEDGEMENTS

Financial support for the production of the first edition of this book was provided by a grant to the Biodiversity Extension Publications Committee from Forest Renewal BC of the Science Council of British Columbia. This committee consisted of Rob Cannings, Don Eastman, David Fraser and Evelyn Hamilton, to whom I express my thanks for their support.

A large number of people helped make this book possible, and I am especially indebted to Ian Hatter, David Nagorsen and Gerry Truscott who were constant sources of information, helpful advice, and encouragement. I also wish to express my sincere gratitude to the following who generously gave their time to review the various species accounts and other major sections, correcting my errors and providing invaluable information: C. Adkins, H. Armleder, C. Barrette, K. Brunt, P. Davidson, R. Demarchi, D. Eastman, J. Elliot, D. Hatler, I. Hatter, J. Hatter, D. Heard, G. Kaiser, G. Keddie, R. Lincoln, B. Mason, B. McLellan, S. McNay, K. Simpson, E. Terry, B. Webster, M. Wood, G. Woods and J. Woods. Others also very kindly provided data and comments: B. Allison, L. Andrusiak, T. Chapman, P. Dielman, O. Dyer, J. Evans, R. Forbes, D. Forsyth, M. Fraker, A. Gaston, M. Gebauer, V. Geist, L. Gyug, A. Harestad, F. Harper, T. Hurd, D. Janz, D. Jones, G. Kaiser, K. Kier, G. Kuzyk, L. Lilley, D. Low, H. Markides, R. Marshall, M. Paquet, K. Parker, T. Poldmaa, D. Richardson, G. Schultz, H. Schwantje, S. Sharpe, J. Shelford, B. Shultz, R.M. Tait, R. Walker, M. Weston, J. Youds and J. Young.

Access to specimens was generously provided by David Nagorsen of the Royal BC Museum, and by C. Adkins of the UBC Cowan Vertebrate Museum. Thanks to M. Tayler and K. Webber for library searches. Rick Pawlas drew the distribution maps based on originals developed by D. Blower – my thanks to both and also to Eric Leinberger, C. Chestnut and D. Salisbury for additional map work. Special thanks to Michael Hames and Denise Koshowski for their excellent illustrations and their patience with my numerous requests.

The original manuscript was reviewed by Alton Harestad, Ian Hatter and Ian McTaggart Cowan, to whom I offer my gratitude for their time and valuable comments that helped improve the work.

Finally, I thank my family and my graduate students for their patience.

Hoofed Mammals of British Columbia

Edited, designed and typeset by Gerry Truscott, RBCM, with assistance from Alex Van Tol.

Typeset in Plantin 10/12 (body) and Optima (captions).

Index by Carol Hamill.

Cover design by Jenny McCleery, RBCM.

Digital reproduction for drawings and maps in this edition by Carlo Mocellin and Shane Lighter, RBCM.

Printed and bound in Canada by Friesens.

THE ROYAL BC MUSEUM

British Columbia is a big land with a unique history. As the province's museum and archives, the Royal BC Museum captures British Columbia's story and shares it with the world. It does so by collecting and preserving millions of artifacts, specimens and documents of provincial significance, and by producing publications, exhibitions and public programs that help to explain what it means to be British Columbian and to define the role this province plays in the world.

The Royal BC Museum administers a unique cultural precinct in the heart of British Columbia's capital city. This site incorporates the Royal BC Museum, the BC Archives, the Netherlands Centennial Carillon, Helmcken House, St Ann's Schoolhouse and Thunderbird Park, which is home to Wawaditła (Mungo Martin House).

Although its buildings are located in Victoria, the Royal BC Museum serves all citizens of the province, wherever they live. It does this by: conducting and supporting field research; lending artifacts, specimens and documents to other institutions; publishing books (like this one) about BC's history and environment; producing travelling exhibitions; delivering a variety of services by phone, mail and e-mail; and providing a vast array of information on its website about its collections and holdings.

From its inception 125 years ago, the Royal BC Museum has been led by people who care passionately about this province and work to fulfil its mission to preserve and share the story of British Columbia.

Find out more about the Royal BC Museum at
www.royalbcmuseum.bc.ca

INDEX